PUT IT ON PAPER !

PUT IT ON PAPER !

Every Person's Guide
to the Printing Industry

MARGIE GALLO DANA

To order additional copies of this book, contact:
Xlibris Corporation
1-888-795-4274
www.Xlibris.com
Orders@Xlibris.com
21427

Contents

DEDICATION

*To my terrific family, Alan and Jacob,
for putting up with my nonstop chatter
about printing all these years.*

About the Author

Margie Gallo Dana, former print buyer turned consultant/writer, is on a mission to break down the barriers between printers and the business community. Through her popular email column, the **Print Tip of the Week,** she shines a light on the printing industry in ways that help print buyers and printers work together better – and have fun doing it. Margie considers herself the printing industry's biggest fan. For more information on her consulting, writing, and public speaking services, visit her web site at www.printconsulting.com or send an email to mdana@printconsulting.com.

Acknowledgments

This book would not have been possible without the many printers, buyers, sales reps, CSRs, press & production personnel, designers, prepress specialists, and paper pros who have helped me learn something new about the graphic arts industry every day of the week. I thank you all.

Please note that I do not work for these sources. I simply appreciate and acknowledge their input, and I'm proud to consider them some of my special "go-to" associates whom I respect and seek out when I need information. I must thank John Elder, my editor, and Linda Powers, who designed the cover (as well as the orginal e-book).

Most of all, I would like to acknowledge all of the professionals who contributed to the columns in this book. I received permission from all quoted sources to use their name and information. Their contact information (company, web site URL, email and/or phone number) was current at the time that this manuscript was prepared for publication.

Introduction

Mark Twain said, "People who love sausage and respect the law should never watch either one being made." I'll take his word on that, if you'll take mine that if you like good printing jobs, you do want to know how they're made. Not just because it's interesting or beautiful (although it is), but because a reasonable involvement on your part will guarantee you a lot better results.

Does your doctor have less and less time to talk with you? Talk to your printers instead! They would love to know more about how you want your print job to look and exactly what you'll be doing with it. They would also love to cut production time and avoid last-minute corrections. Conversations like this can save you money, save you and your printer time and aggravation, and even give you the thrill of a better print job than you expected.

If you're a newcomer to print buying, now is the time to learn how to work with your printer.

Printing—even producing a simple (hah!) brochure—is quite a complicated process. There are many kinds of paper, many kinds of ink, many kinds of printing, and many kinds of print shop. You don't have to know all about them. You just have to know that a good printer does know all about them. But even a good printer can't read your mind, in which case a lot of your printer's knowledge and expertise isn't doing you any good. So if this book had to be boiled down to one word, it would be communication.

Before I became a world-famous print consultant (I can dream, can't I?), I had a career as a corporate print buyer. (We'll leave my cocktail waitress days out of it.) I was never a printer and I never worked for a printer. But I was always interested in printing and the more I learned about what printers can do and how happy they would be to do it, the more convinced I became that all print buyers should see what I was seeing. My goal ever since has been to shine a light on the printing industry in ways that help buyers and printers work together better.

The printing field changes constantly—that's one reason I find it so interesting. Everything in this book is as up-to-date as I could make it. You'll see that I sometimes hand my soapbox over to experts who gave me their permission to tell you where they work, along with a URL or phone number. Some of these great people may have changed companies since the book was compiled.

If you like what you're learning in this book, these tips are only the tip of the iceberg. Visit my website (www.printconsulting.com/Subscribe.htm) and subscribe to my **Print Tip of the Week**. Each week you'll receive a new Tip via email. Hopefully, you'll be a bit more enlightened—and excited—about working with the printing industry because of it.

I hope you enjoy this book. It's my first.

Margie Gallo Dana
www.printconsulting.com

Chapter 1

Tips for Beginners

Professional print buyers start out knowing very little about printers and printing.

There are no college degrees in print buying, and very few educational institutions offer any kind of training in this specialty field. It's something you learn on the job.

So for all of you who are about to buy printing services for the first time, take comfort in knowing that we all started out in the same place.

This chapter provides some very basic information to help you get it done right the first time.

Common-Sense Advice for the
First-Time Print Buyer

Not all printers are created equal. Most are general, commercial printers, rather than specialty shops. Lots of consumers assume that all printers do the same thing. It's not that simple. Different presses mean different capabilities and different run lengths. You want a "good fit."

Ask someone you know for the name of a printer he or

she has used successfully. This printer may not be exactly right for your job, but it's a smart start. If you search for a printer on your own, you may or may not get lucky the first time. Printers come in all sizes.

Want three estimates? Expect three different prices from three different printers. They should be in the same ballpark, though, if you've chosen well. Pick the printer with whom you feel most comfortable.

Always ask for prices before you hand over your job to a printer. Printing prices vary from shop to shop. Exceptions: prices are often posted in a retail print/copy shop. Just know the cost before you have the job done. The point is to avoid surprises.

Paper accounts for about 30% of a print job. Sometimes it's higher; sometimes, lower. Select your paper carefully with the printer's help. If you're quoted a price that is too high for your budget, ask your printer to suggest some alternatives.

Determine your quantities wisely. It's usually more expensive to go back on press to print more copies than it is to print more the first time.

One-color jobs, using black ink, are the cheapest. But color sells, so get prices for 2-color and even 4-color versions if they're appropriate for your job. Prices for color printing have come way down.

Remember, a 2-color job is not twice the price of black ink only. Similarly, 4-color printing doesn't mean it's twice the price of 2-color printing.

Make your deadline known to the printer. Don't use vague terms. You can't expect the printer to know what "ASAP" means to you.

Proofreading is your job, not the printer's. Check and re-check your copy before handing it over to be printed. The earlier you catch mistakes, the cheaper they are to fix.

Always ask to see a proof. Even on reprints, no matter how minor your type changes are. You're more likely to make mistakes when you're in a hurry to redo a job. Discuss it with your printer and find out when the proof will be ready, what you're supposed to do with it, and how fast you need to turn it around.

Printers accept most jobs digitally these days—that is, directly from your Mac or PC, or from floppy disks, Zip disks, or CDs.

What you see on your computer monitor is not what you'll get.

Most of the time, your files are not absolutely ready to print. The printer will need to clean them up—and is entitled to charge you for it!

Word files cannot be converted to 2-color offset printing jobs yet. They can only be printed in 2 colors on a digital printer or laser copier

When ordering business cards, don't stop there! Think about letterhead, envelopes, mailing labels, and blank sheets for those 2-page letters of yours.

Don't assume your printer will save your logos, original files, etc. Get your files back from the printer when your job is done, so you have them on hand when you need reprints in a hurry. And you will.

Once you give a job to a printer, the work has just begun, and it will take days or weeks (sometimes longer!) to produce it. Don't expect miracles.

Be mature enough to admit you know very little about printing. Ask for help if you have questions! Printers aren't mind readers, and they want your experience with them to be good so you'll keep coming back. I'd encourage all new print customers to say, "I've never worked with a printer before—where do I start?"

Parlez-Vous Printing?

Is there a Berlitz Guide to printing? There should be. Printers have a language all their own.

If you plan to be in the print-buying business for any length of time, you'll have to wade through an alphabet soup of printing terminology that's often confusing. This language barrier is probably the #1 reason for dissatisfied customers— they didn't know what they were asking for!

There are hundreds of technical terms about typesetting, design, printing techniques, inks, and binding options. Don't get me started on paper: there are different grades, different characteristics, different brand names, and different weights, sizes, thicknesses, colors, and finishes. The terms keep changing along with the technology.

You can find decent glossaries in most of the trade books, including *Pocket Pal* by International Paper and *Getting It Printed* by Mark Beach and Eric Kenly.

Here are 24 key terms that every new customer should know.

AA's & PE's If you make changes to a job at the proof stage, they're either AA's or PE's. Author's Alterations are changes that the customer makes. You should expect to pay for AA's. Printer's Errors are mistakes the printer makes. You don't pay for PE's.

Bleed Describes ink coverage that goes right to the edge of your paper, so the color "runs right off the sheet." Can increase your cost, so tell your printer up front.

CMYK and RGB Computers and printing presses use two different color systems. CMYK refers to the 4 colors used for full-color process printing: C for cyan (blue), M for magenta (red), Y for yellow, and K for black. Full-color images, like photos and illustrations, consist of patterns of these 4 colors.

Computers use RGB, which stands for red, green, and blue. Color on a computer monitor consists of millions of dots in these three colors.

CSR Customer service reps, the key behind-the-scenes

employees in a printing firm. Good ones know your jobs intimately. Treat 'em right!

Digital printing The "newest kid on the block" when it comes to printing methods. It's such a widely used term in the industry these days that 10 people will give you 10 different definitions. In the broadest sense, it describes printing from electronic files directly to a press—without going to film. Made for short-run, fast-turnaround printing, from one color to full color.

Full-color or 4-color process printing The use of 4 specific colors—cyan (blue), magenta (red), yellow, and black—to reproduce color photos or illustrations. More expensive than 2-color printing.

1-color, 2-color, 4-color These terms indicate the number of inks in a print job. Normally, 1-color means black ink, and a 2-color job means black plus a spot color. A 4-color job most often means using the 4 process colors (CMYK), but it could mean 4 spot colors, too.

2/2, 2/1, 4/0, etc. Shorthand for 2 colors both sides; 2 colors on one side, 1 color on the other; 4 colors on one side, no printing on the other. So the next time a printer says something like, "Does this print 2 over 1 or 2 over 2?" you'll know what he means!

1M, 2M, etc. Shorthand for quantities you need printed: 1,000, 2,000, etc.

50#, 70#, etc. Describes the weight of paper. # is a symbol for pound. Let your printer or designer guide you when selecting the best paper weight for your job.

Halftone A reproduction of a photograph, usually a black and white photo.

Makeready Describes all of the steps that are required to set up or prepare a printing press before printing a book, magazine, or indeed, any printed product.

Offset It means two things. First, offset lithography is the most popular kind of commercial printing method, whereby ink is transferred or offset from a printing plate to a rubber cylinder and then to another cylinder that holds the paper. The other meaning: when printed sheets are stacked on top of one another as they come off the press, wet ink can "offset" from the top of one sheet to the bottom of another. Also called "setoff."

PMS Stands for Pantone Matching System, the standard ink color system used by printers. Ask to see a PMS swatch book when choosing ink colors. There are thousands of colors to choose from—which is why your request to print your letterhead in "black and green" is nowhere near specific enough.

Prepress Describes the necessary steps a printer takes to prepare the customer's application files (such as QuarkXpress, PageMaker, Photoshop, etc.) to be imaged at high resolution to produce printing plates. Basic prepress steps include RIPping, trapping, and imposition. (See "The Scoop on Prepress" in this book.)

Process color The same thing as 4-color or full-color.

Reverses or KO's (knock-outs) Describes type or images that reverse out of a surrounding area. Picture a solid black band (or any color) that has white type or a graphic inside of it. This white type or image is a reverse or a KO.

Saddle stitch or saddle wire A common binding method, whereby the spine of your booklet is bound by two staples at the fold.

Scoring Creasing a sheet of paper so that it folds more easily and won't crack. Scoring is done at the printer's by machine, not by hand.

Sheetfed printing Sheetfed presses use paper that's been precut into sheets.

Specs Short for specifications. Specs are all of the details that describe your job exactly, allowing the printer to provide a fair estimate. They include the dimensions of your piece, the quantity, ink colors, paper, and so on. Writing complete, accurate specs is key.

Spot colors Ink color choices that aren't 4-color (which is process color). Most often you'll choose colors from a PMS (Pantone Matching System) swatch book, available at your printer's. Also known in the trade as "flat colors."

TIFF, JPEG, EPS & PDF The most common file formats that graphic designers use to send bitmap photographic images to printers and clients. TIFF stands for Tagged Image File Format; JPEG stands for Joint Photographic Expert Group; EPS stands for Encapsulated PostScript; and PDF stands for Portable Document Format. They're recognized on both Mac and PC platforms. They have the .TIF, .JPG, EPS, or .PDF extensions in the file name.

Web printing Web offset presses use paper that comes in webs or rolls. In general, web printing is used for very long runs. This term has nothing to do with the Worldwide Web!

The Major Types of Printing: A Primer

"I'm a printer" can mean many different things, depending on the equipment that a printer uses. Different printing processes produce results that look and sometimes even feel different to the touch. Here's a brief description of the major types of printing processes.

Offset Lithography The most common printing process today. Offsets ink from metal plates to a rubber blanket (cylinder) to the paper. Almost every commercial printer does offset printing.

Letterpress The original process created by Gutenberg

around 1440. As with rubber stamps, images on the plate are higher than the surface. Fine letterpress is being done by fewer and fewer printers.

Flexography A special type of printing for packaging products. The plates used are flexible. Products include cardboard boxes, grocery bags, gift wrap, and can and bottle labels.

Gravure Prints directly from cylinder to paper. Used when printing for millions of impressions—think magazines, newspapers, and direct mail catalogs.

Screen Also known as silkscreening. Ink is forced through a screen, following a stencil pattern. Used for ring binders, T shirts, bumper stickers, and billboards.

Engraving Think "fine stationery." This process produces the sharpest image of all. An engraved image feels indented (run your fingers over the back side of the sheet). Many law firms still use engraving.

Thermography A raised printing process that's less expensive than engraving. Thermography uses special powder that adheres to any color ink. It's mainly used for stationery products.

Reprographics A general term covering copying and duplicating. Think in-house copying departments and copy or quick-printing shops. They take your originals and make duplicates of them.

Digital Printing The newest printing process and the least understood! The term includes all processes that use digital imaging to create printed pieces. Doesn't use film. (Think desktop to the digital press.) For short-run, fast-turnaround jobs.

The Paper Trail—from Mill to Merchant to Printer

When I was a corporate print buyer, I was intrigued by the winding road that paper took to get to printers. Today, clients still ask a lot of questions about paper, confirming the ongoing confusion about the paper trail. Here's just a snapshot of that route.

Paper comes from pulp, which is converted into paper by PAPER MANUFACTURERS. There are several dozen major paper manufacturers in North America. These papermakers typically own several mills, each being devoted to a certain kind of paper, such as coated or uncoated. Every paper is given a specific brand name by the mill.

PAPER MILLS sell their paper to MERCHANTS. Different merchants can carry the same brand of paper. Some merchants have "private label" paper, which they name themselves. Note that some of the larger papermakers also own merchants. Are you still with me?

Paper mills and paper merchants both have paper specialists. Called SPECIFICATION CONSULTANTS, their job is to meet with buyers/graphic designers and recommend papers, discuss trends, and supply dummies and swatch books (paper samples).

Merchants sell paper to printers. Printers buy their paper from several merchants, since no one merchant carries every brand. Some printers buy paper directly from paper manufacturers, too.

If you're a consumer, you should let your printer order your paper for you, with a few exceptions:

— A very small business owner who wants, let's say, a ream of letterhead and envelopes, can pick up paper at a local office supply store. You won't have many options, but in a pinch, it will do.

— Large-volume corporate print buyers sometimes buy their paper directly from a merchant. These folks really have to know their stuff to do it successfully.

When it comes to paper, this is just a Tip—not the iceberg.

4-Color Process in Plain English

"Do you want to print this in 4-color?"

"No, I want to use all the colors."

This verbal exchange between a print consultant and a client really happened, and it points out a very common misconception about 4-color printing.

More difficult than other types of printing, 4-color printing refers to the process of using 4 specific colors in combination. These colors are cyan (blue), magenta (red), yellow, and black. (You'll often see them abbreviated as CMYK.) These 4 colors are always used.

4-color printing is used for very specific reasons: to reproduce color photographs and illustrations. The colors are used in combination to simulate the desired hues in a particular design. They're actually layered on top of one another. If you look at a printed image of a full-color photo with a magnifying glass, you'll see a pattern of blue, red, yellow, and black dots—not solid colors.

Unless your design includes color photos or illustrations, you most likely won't need 4-color printing. Go with spot (or "flat") colors instead. And remember, not every printer can do 4-color printing.

Chapter 2

CHOOSING—AND KEEPING—
A PRINTER

One of the trickiest things about print buying is finding the right printer (or printers). You'd think there would be an easy way, but there isn't. There are about 45,000 printing companies in the U.S., most of which are independently owned and operated. So finding the right printer is somewhat of a hit-or-miss situation.

Once you find a great printer to work with—one who serves you well and delivers great quality—there are ways to build on that business relationship.

This chapter deals with these key issues of finding and building the right business relationship.

Not All Printers Are Created Equal

Many people think that all printers do the same thing. Think again. For example, some specialize in annual reports, some in stationery products, and some in magazines. Some do copying as opposed to offset printing. Some have only sheet-fed presses, ideal for shorter runs; others have web presses, used for longer runs. Some have desktop publishing expertise. Many don't.

And not every printer does 4-color printing.

How do consumers find the appropriate printer? It's not easy. A lot depends on the printing equipment that a print shop has. You can ask to see an equipment list—but these lists are so technical that they're virtually useless to the average print customer.

I recommend you interview printers about the kinds of jobs you need printed. Get samples of similar work they've produced for other customers. Ask lots of questions, so the salesperson is clear about the kinds of print jobs you need and the service you expect. And it's always a good idea to get estimates from a few vendors.

Choosing the right printer is a lot harder—and makes a lot more difference—than you think.

"You Can't Buy Service"

Unless you absolutely have to buy "price" when dealing with printers, don't focus on it.

If you choose your printers wisely, their prices will be in the same ballpark anyway. You'll have preselected some printers who have the right equipment for your particular products. The quality should be there already, presuming you've seen samples and/or have some solid references from trusted colleagues.

"You can't buy service," one professional print buyer told me. She selects vendors based on how they handle her jobs and how her rep treats her. She also likes going on plant tours and seeing clean print shops. When she does a press OK, she likes being right there next to the press, not tucked away in a private "viewing" room.

So when you're shopping around for a new commercial

printer, pay attention to the service. How are you treated when you call? How quickly does the sales or service rep return your call? Do you get prompt and accurate information about your print job when you request it? Is the phone system a pleasure to use?

No matter what's happening with the economy, expert print buyers tend to shop for service providers, not bargain printers. You want my advice? Take theirs!

With Printing, Do Actions Speak Louder than Words? Yes . . . and No

The best printers are not just interested in making a sale. They're interested in finding solutions for you. They treat your business with respect. You'll see it in their actions; you'll hear it in their words.

Here are some of those actions:

Exquisitely printed samples of jobs that mean something to you. Samples of things you're likely to print.

The plant tour. Many buyers have told me that they really like seeing a printing plant, so that they can see the equipment firsthand and make sure the plant is clean.

A demonstration of the salesperson's accountability to you. How accessible is the sales rep (and/or CSR) when you need information? How fast and how professionally does your sales rep respond when there's a problem?

Delivery of materials. Is a print job delivered when and where it's expected? Does it look like you expect it to look? Without a doubt, this is what printing is all about!

But with all the competition among printers, there are also words that print buyers should listen for when they're contemplating which printers will make for successful, long-term relationships. Here are some of those words:

"Tell me, have you ever worked with printers before? What was that like?"

"I'd love to send you samples that mean something to you. What kinds of things do you print, so that I can do that?"

"Do you have any questions about how we print something? Is there anything we've discussed that you don't understand?"

"I hope to work with you for a long time. Is there anything you can tell me about how you like to work with printers that will help our business relationship?"

"I blew that deadline—I'm sorry. I'll do everything to make sure it doesn't happen again."

"It's year end, and you've been a great customer. Tell me, are we meeting your expectations, or is there something we should be doing better?"

"Thanks so much for the business this year. It means a lot to us."

20 Questions—
or How I'd Interview a Print Salesperson

When I was a new print buyer, I don't recall interviewing many printers, because I inherited most of them. Even when I met new salespeople, I was too "green" to know what to ask!

If I had to interview a printing salesperson today, here are 20 things I'd ask:

1. Tell me about your firm's history and philosophy.
2. How much industry experience do you have, and in what areas?
3. With so much competition in printing, what makes your company different from other commercial printers?
4. How many shifts do you run? In prepress, too?
5. Do you do mainly 1- and 2-color work, or full-color? What other services do you offer? How about mailing and fulfillment?
6. My company prints a lot of X, Y, and Z. Do you print much of these products? What's a typical quantity for you? Can I see samples?
7. What kinds of processes do you typically outsource?
8. Do you have a graphic design department? How big is it? Do the staff members do actual design or are they primarily desktop publishers?
9. What software programs can you support?
10. Can you handle PC files or only Mac files?
11. Will you be my primary contact? Will a CSR (Customer Service Rep) be assigned to me as well? How long has he/she been with the firm?
12. How do I send my jobs to you?
13. Do you have an Internet site, with FTP capabilities?
14. How do you keep in touch with your clients?
15. Can you please give me the names and numbers of three current customers who have the same type of work that I would give you?
16. Do you have formal QC programs at the plant? How do they work?
17. I hear that the industry's changed a lot lately—how

is your firm keeping up with new printing technology?

18. Other buyers on my staff are quite new at this; how can you help?

19. Some of our content is already up on our firm's Web site. Can you help us figure out how to make it print-ready?

20. Do you offer print management tools via the Internet? Can I access my jobs online, see their status, and check my inventory? Do you have online bid forms and production schedules?

These questions don't address personality (do you like this person?), quality (how do those samples look to you?), or price (you'll have to obtain bids). But the answers will tell you the kinds of services a company can offer you—and how much this salesperson knows about the company and the industry.

Look Quick!

Printers Are Doing Everything!

The term "quick printer" is disappearing.

While reading an issue of *Quick Printing* magazine, I realized that many consumers probably don't categorize printers as "quick" or "commercial," "financial" or "digital." To many people, a printer is a printer is a printer.

That got me thinking about the original concept of a quick printer. I asked Michael Vogel, owner of Sir Speedy Printing in Milford, Connecticut, to define the term for me.

"Quick printer" evolved from "instant printer," a term that originated in the '60s, when such shops produced short runs on duplicator presses using the new Itek paper plates rather than traditional metal ones. These new plates could be made in about a minute for a fraction of a dollar.

Taking advantage of the new technology, this new breed of entrepreneur could produce printed documents in a few minutes—fast enough so customers could wait for them. Commercial printers at that time typically offered turnaround times measured in weeks. "Quick printers" brought a customer-oriented attitude to the craft by coupling new technology to a new mindset.

The new mindset was given a boost when, in the late '80s, the popularity of fax machines coupled with corporate adoption of FedEx "gave new meaning to instant communications," noted Vogel. Suddenly, consumers expected everything instantly. Burdened by recession-caused overcapacity, many printers were reluctant to say no.

At about the same time, Canon introduced the first reliable color copier, the CLC 1, which became the darling of the quick-print industry. Then came the desktop publishing phenomenon, which created a demand for short-run color jobs. Quick printers started adding more offset equipment to handle this work. Thanks to enabling technologies, they could afford to do process color printing in-house rather than outsource it to commercial printers.

Until this time, "quick printers" weren't totally accepted by sophisticated buyers. But as technologies improved and print runs shrank, that attitude changed. Plus, today's crop of buyers doesn't have the same prejudices.

In fact, today's buyers often think of the "corner print shop" as the only print shop they need, because these printers can fill most needs, whether they do it themselves or outsource it.

And larger commercial printers have been investing in high-end digital copiers, such as the Xerox Docutech, for years, because of the exploding demand for such work.

So you see how the line between quick and commercial printers is blurring. Test it out: the next time you're with some business people who aren't professional print buyers, ask them where they'd go to print their color brochures, or direct mail pieces, or corporate stationery. Chances are good that they'll name their local quick printer.

Scorecard #1:

How Well Is Your Printer Serving You?

Here's a questionnaire for you to answer, to guide you in deciding how well you are being served by the printer—or printers—with whom you work.

PRINTER SCORECARD

- Do I get accountability and reliability?
- Do the salespeople hold up their end, see the job all the way through delivery, and take responsibility for troubleshooting?
- Does anyone engage in finger-pointing of any kind?
- Do I have one conduit with a company, be it the salesperson or the CSR? It's important to know one person has a handle on my work.
- Does the printer do everything possible to keep to an agreed-upon schedule—or to forewarn me if the date's in jeopardy?
- Does the printer keep me updated on the latest printing trends and technology?
- Does my sales rep ask me a lot of questions?
- Does my sales rep have a great rapport with the pressmen in the plant?
- Can I easily communicate with my sales rep in person or with email? This relationship is key.
- Is this printer honest and direct? Do they take jobs they know they can't deliver on time?
- Does this salesperson know much about my business?

- Am I getting service, quality, and price—in that order?
- Do I get premium quality when that's my highest priority?
- Will this printer develop into a long-term partner? I don't like shopping around if I can avoid it.

Scorecard #2:

How Well Is Your Sales Rep Serving You?

For most print buyers, a printing company is personified by the sales rep. Talk about pressure to perform!

Printing sales reps are judged by a variety of criteria, depending on the individual print buyer.

Here are six obvious qualities to look for in a printing salesman or saleswoman:

SALES REP SCORECARD

1. Mutual comfort level (good, old-fashioned rapport).
2. Industry knowledge and experience.
3. A very high level of service, including responsiveness, availability, and commitment to delivering jobs on time.
4. Honesty.
5. Dependability.
6. Self-confidence.

But four additional, less obvious qualities are equally important:

1. Keeping you up-to-date with industry trends, especially the ones that will likely affect you.
2. Recommending printing alternatives that will save you time, money, or both.

3. A willingness to accept responsibility when things go wrong.
4. A demonstrated, genuine rapport with the pressroom employees, indicating a good working relationship.

Chapter 3

PREPARING YOUR MATERIALS

"The time has come, the walrus said, to talk of many things." Of fonts and Macs and PDFs, of color printing stings.

But let's start with something non-technical and absolutely paramount, the very essence of preparing materials for the printer: Proofreading.

Never Proofread Your Own Stiff

Uh, sorry, make that "stuff." This is a very old rule of thumb that bears repeating on a regular basis. You may be a keyboard wizard, but if you're creating content for a printed (or Web) piece, make sure that someone else proofreads what you've done.

Because very often, you can't see your own mistakes.

Pay particular attention to bits of copy that you tend to type over and over again, like your business address, phone number, email address—even your own name! In my many years as a proofreader, I can't tell you how many times I saw "Untied States of America" and "works in the pubic domain."

Embarrassment aside, if some typos are egregious enough, you'll have to reprint the job and pay for it twice.

To prevent printed bloopers, establish a procedure whereby someone other than the desktop publisher or graphic designer is designated to read every bit of copy. Supply your "Designated Proofer" with a checklist for names, dates, prices, and other key information.

If you or your designer creates a digital file for a print job, it's your responsibility, not the printer's, to proofread. Even if your designer does the keyboarding, he or she will likely ask you to sign off on a form that says you've OK'd the copy for printing.

Hire a freelancer if you must, but do not skimp on this step. Don't regard spell checking as a substitute for proofreading, either. It's far from perfect.

Did you know that there are traditional proofreaders' marks—a kind of shorthand that's known by printers and designers alike? Check them out at this link on the Merriam-Webster site: www.m-w.com/mw/table/proofrea.htm, especially if your documents are lengthy. I still use them to this day.

P.S. Proofread earlier rather than later in the process. You want to avoid finding typos on press. They'll cost you plenty to fix: plenty of time and plenty of money!

Creating Your Own Files?
Avoid These 10 Common Problems

Having an important print job done without using any electronic files is like going to the bank and getting cash from a teller. You can do it—but I'll bet you don't do it very often.

As in just about every other part of life, computers make many new things possible, including many mistakes and pitfalls.

As a member of the PrintShare List (check out www.printweb.org for list information), I get a bird's-eye view of a range of issues faced by printers across the country. A key discussion focused on the most common problems print shops have with customer files.

Here is a compilation of those common file problems. As you're preparing files for a print shop, pay attention to the following:

1. FONTS, FONTS, FONTS. Sometimes they're not included with the document. PC fonts on a Mac cause problems. Customers "use every non-traditional script typeface in all caps" then wonder why the document isn't readable at 100%. People use slightly hybrid versions of a traditional typeface that create havoc, too.

   Mixing TrueType and Postscript fonts cause problems, because TrueType fonts are single file fonts that contain data for both low-resolution screen rendering as well as high-resolution output. By comparison, Postscript fonts only contain the high-resolution data. When imaging at high resolution for printing, the RIP (Raster Image Processor) utilizes the font data. Since RIPs are Postscript-based as well, they communicate with Postscript fonts better than with TrueType fonts.

2. Lack of knowledge on the customer's part. "What software did you use to create this file?" "Windows." Trying to explain spot separation to someone using process clip art.

3. Image resolution too low (72 ppi) or in the wrong format (JPG). Customers wanting low-res scans to print to high quality because "it looks OK on the computer screen at home!"

4. Designing solids that cause ghosting.

5. Specifying process color for text instead of black.
6. Trapping spot colors.
7. Customers wanting printers to match the color off of the free inkjet printer they got with their computer.
8. Hidden copy in text boxes.
9. Not including a hard copy of their file when they send it in.
10. Did I mention fonts?

Customers who create files for printers can help avoid these problems by talking with their printers early. Ask for tips on file preparation.

Remember, just because you can use a computer doesn't mean you can design a job to be printed commercially. As printing expert Frank Romano once said, "Anyone can own a violin—but not everyone can play one."

PDF: The Remarkable, Versatile Digital Proof

If you're not using PDFs to communicate with your designer, printer, and others who need to see your files as you've created them, get with the program!

A PDF stands for Portable Document Format. It's created using Adobe® Acrobat®.

The beauty of a PDF is that it takes a snapshot of your document exactly as you created it. Then you can send it digitally to anyone else who needs to see it, even if the other person doesn't have the program you created it in.

A PDF file preserves your fonts, your layout, your graphics. So what you see on your computer monitor is exactly what's seen by anyone who opens your file.

To create a PDF, you (or your designer) must have the Adobe Acrobat software. But anyone can read a PDF file, because Adobe® Reader® can be distributed freely by the person who created the PDF file or downloaded for free over the Internet.

Four great uses for PDF files:

1. As email attachments to your printer to get an estimate.
2. To communicate with your designer digitally.
3. To share your files with others—down the hall or clear across the country—who need to see/approve a design.
4. To send a document you've created to people who don't have the program you created it in.

Faxes and paper proofs pale in comparison with the PDF. So look into it PDQ. And save yourself time, money, and frustration.

Not So Fast, Lady. Macs Don't Always Rule!

"The platform of choice among printers is still the Mac. Programs like Word and Publisher were not created for output on a commercial press," I wrote in one of my Print Tips.

"This may be true of some printers," wrote Roy Nix of Nix On Time Printing (www.nixontimeprinting.com and www.fastcolorprinting.com), "but most good shops today will take both platforms and never indicate which they prefer. Any printer who uses Macs exclusively is in danger of losing market share each day."

He's right. While graphic designers and other print professionals might still prefer Macs—even turn their noses up at PC programs—more businesses are buying PCs than Macs, and printers are becoming more "ambidextrous," shall we say?

The topic of PCs vs. Macs always comes up when I'm interviewing printers. Little by little the PCs are gaining ground in print shops, and while the majority of printers have said that they work with Mac files as a rule, PC files have to be dealt with (and printed from).

A printer from California echoed Roy's comments. "Most printers I know support both platforms and many prefer PC. We're primarily a PC shop because our clients are businesses, and businesses run PCs."

John Giles wrote a column in Quick Printing magazine (www.quickprinting.com) about Publisher 2002 and how this new version has greatly expanded its commercial printing features. "With millions of users, Publisher is becoming a force in the quick printing marketplace," he writes. In fact, Microsoft offers free software and support to printers and service bureaus that want to support this software. (Details are available at www.microsoft.com/publisher/pspp/.)

"I have watched the transition from being anti-Publisher to being Publisher providers," noted Harry Silvis of Kennesaw Graphics in Georgia. "These printers have realized that many customers are not going to purchase Quark or PageMaker and that they needed to be able to deal with Publisher." He also said that Word has been a very profitable program for his shop.

I don't think that Word and Publisher files are always easy to work with. It certainly depends on the creator and the printer. (You could say the same thing about Mac programs, too.) But it's an ongoing issue and I'm happy to stand corrected.

The times they certainly are a-changing.

Type Tips

Everyone has a personal computer these days—and suddenly, everyone is a designer! But using type appropriately isn't as simple as selecting different fonts and sizes from your pull-down menu.

Try to avoid these four common mistakes with type:

1. Using sans serif type for body text (makes it too hard to read).
2. Mixing too many typefaces (and sizes) in one document (unless, of course, you're typing a ransom note).
3. Having lines of text that are either too long or too short (hard on the eyes).
4. Using large blocks of capital letters in copy (takes too long to read).

If you can't afford to hire a graphic designer, think "less is more" when using typefaces.

File Formats for Graphic Images

Ken Hablow of KH Design (www.khgraphics.com) in Weston, Massachusetts, provided this guide to common file formats for graphic images:

There are several file formats that graphic designers use to send bitmap photographic images to printers and customers. The most commonly used image formats are TIFF (Tagged Image File Format), JPEG (Joint Photographic Expert Group), EPS (Encapsulated PostScript), and PDF (Portable Document Format).

They're recognized on both Mac and PC platforms. You've seen them: they have the .TIF, .JPG, .EPS, or .PDF extensions in the file name.

When saving these files for print output, think about resolution. Normal computer screen resolution is 72 dpi (dots per inch) for a PC and 96 for a Mac. The rule of thumb is to save an image for print at double the line screen of the final piece. If you're creating negatives with a line screen of 150 lpi (lines per inch, so that's 150 rows of dots per inch), save the image at 300 dpi.

TIFF files (TIF on PCs) form high-resolution images. They're large and often sent uncompressed. JPEG files are compressed to save transmission speed. But talk to your printer first about image files—some prefer that you don't compress them. Each level of compression deletes some dots, so the quality suffers.

An EPS file should be saved in the exact format for the specific platform (PC or Mac), and sometimes even for the specific program that will be used to open it. EPS files don't always transfer from PC to Mac very well.

Here's a good tip: if you're a Mac user and want to send an image file to a PC user, save the file with the exact file extension. PC programs look for the extension first. Mac programs are more tolerant of files with no extensions.

PDF files are another format choice. Some printers and service bureaus can color separate from PDFs, but check first.

The moral of this story is that the design part of digital prepress is incredibly technical. Don't make the "file transfer moment" the first time you connect with your print professional. Do it much earlier to be sure you are preparing files that are printable. It's not kid's play; it's complicated stuff.

Printing from Word Files?
Think One Color

Attention, all you folks who work in Word or PowerPoint: Before you hand your files to a commercial printer, you need to know that Word and PowerPoint files cannot be color separated.

This means that no matter how "colorful" you make them look on your desktop, these files can only be output to film in one color. Then they can be hand-stripped to multiple colors. That's expensive.

However, these same files can be output to digital color copiers or black and white digital copiers.

I recently created a 2-color flyer on my PC in Word. I output it to a Canon Laser Copier so I could get it printed in 2 colors ASAP. Is it "high art"? No. But it's perfect for my needs. An offset printer would have had to hand-strip the film from my files in order to give me 2 colors.

Specialists in preflight technology are working on correcting these file issues, so that both customers and printers will have an easier time getting the results they want.

What can be done in the meantime? Customers should check with a printer before preparing Word or PowerPoint files, just to be sure he or she can deliver the expected results.

Too bad a pop-up message doesn't appear when you begin a new Word document: "Planning to bring this file to a print shop? Please call your printer first."

Color Your World RGB—or CMYK??

Converting color images into files that can be used on a printing press is very, very complicated. Computers and printing presses use two different color systems. Computers use RGB and presses use CMYK.

And RGB has to be converted to CMYK before a job prints.

RGB stands for red, green, and blue. Color on a computer monitor consists of millions of dots in these three colors.

CMYK refers to the four colors used for full-color process printing: C for cyan (blue), M for magenta (red), Y for yellow, and K for black. Full-color images in printing, like photos and illustrations, consist of patterns of these four colors.

"Who should convert the files—the designer or the

printer?" I was asked recently. I checked with trusted colleagues, both designers and printers. The answer? "It depends."

Whoever does the RGB-to-CMYK conversion can control how the color separations are built.

Designers experienced with printing may prefer to convert RGB to CMYK, using different image-processing software. But this isn't like saving a file with a different name. It's more like telling a software program to build color separations, and that isn't a straightforward process. Sometimes designers don't know exactly how their files will be converted—so they won't have any idea what their images will look like in print until they see a proof, which is far too late in the process.

Some designers send in RGB files, knowing that their printer will do the conversion expertly. Other designers say that it depends on the images and where they came from. Sometimes they bring images into their designs that have already been separated into CMYK.

Keep in mind that colors are device-dependent, so what you see on your monitor differs from what I might see on mine—and so on and so on.

Frank Romano sums it all up in his *Pocket Guide to Digital Prepress*, where he wrote: "What the scanner sees, the monitor shows, the proofer proofs, the film images, and the press prints are essentially different."

If you're a designer, call your printer to discuss RGB-to-CMYK conversion.

When Color Isn't WYSIWYG

If you're creating files to send to a commercial printer, you'd better know how to do it right. Creating good, clean printable files that translate into the colors you expect to see is neither simple nor a one-step process.

Ken Hablow, principal of KH Design (www.khgraphics.com) in Weston, Massachusetts, had the following experience:

"I had a photographic image I needed to screen to 10% as a full background and have color separation negatives made for magazine display advertising. The separations are CMYK. Of course, my monitor is RGB, and so was the original photo. I did all the right things—almost. The photo was purchased as a royalty-free image. Knowing I had to ultimately output this as CMYK, I first converted the image to CMYK in my paint program, manipulated the image to the 10% I needed for the background, and imported the CMYK version into my layout program. It looked great on screen, and the client was happy with the color.

But the computers at the service bureau were not.

The image was a concrete building against a deep blue sky. When screened down to 10%, the building went almost white, which is what we wanted, and the sky to baby blue—on my RGB monitor. The CMYK imagesetters and composite printers all output the baby blue as a bland green. I tried PDF files with different settings, and also direct PRN files, for both the separations and the color matching composites. They all output as green.

My service bureau really takes the word "service" seriously. Their best tech person took the image apart and discovered there was about a 5% difference in the amount of cyan, magenta, and yellow in the sky. At full saturation this would not be a problem, but at 10% this meant there was half as much yellow as cyan; and we all know blue and yellow make green.

It took a lot of work at the service bureau to eliminate all traces of yellow and magenta before the image printed properly, but the final match print was exactly what my client had seen on my monitor.

Another problem was that my service bureau uses a different color-matching profile than I do, which can alter the color hue and saturation of an image. So, in the future, when I am screening photographic images, I will have a match print made before I start assembling the page.

This was an expensive lesson as we output three full sets of negatives for the three different magazines before we got it right, several color matching prints, and two versions of the ad as a large trade show display. So, the moral of this experience is: work closely with your service bureau, run a match print, and be sure that, before you go to press, you have the color right. What you see on your monitor will probably not match the finished product."

Designers beware.

Chapter 4

Placing Your Order

Printing is a manufacturing specialty that requires lots of input and information from you, the customer. You can't just drop off a job at the printer's . . . not if you want great results. Your printer needs lots of information about you and your project. In a sense, you're building it together. The smartest thing you can do is talk with your printer even before you start designing your next print job.

This chapter will help you appreciate why.

A Printer Is Not a Mind Reader

A lot of clients express frustration over print jobs that didn't look like they'd imagined. Often—not always—this happened because they never really told their printer what they wanted.

Whether you're handing the printer a Mac file that's been prepared by your graphic designer or a stack of hard-copy originals from which to print, you should always have a conversation with your printer about what outcomes you expect.

Talk with your printer early and often. Provide dummies

and/or laser proofs of your job, complete with color breaks and folds, if appropriate. A dummy, or handmade layout, will show the printer how your piece should look when it's printed. The best printers will study the dummy and alert you to potential problems before they print the job. These are the printers you'll want to use for the long term!

Strive to make your printer your business partner, not merely a vendor. Don't assume that he or she knows what you're envisioning. With printing, your involvement and input definitely influence the finished product. So the more you communicate, the more likely you'll get what you want.

Estimate This!

I love it when a client asks, "How much will it cost to print my brochure . . . ballpark?"

That's like asking a builder, "How much for a house?"

OK, you're not about to invest $350,000 in a new brochure, but you get the picture. In order to get a good printing estimate, you first have to provide good specs (specifications).

In fact, it's often wise to seek estimates from three or four printers, just to be sure you're getting a fair price.

First, write really good specs, which define your piece in precise details to a printer (see next article). Key details include paper stock/weight/color, quantity, number of inks, finished size, page count, binding method, etc. In printing, details matter, so the tighter the specs, the better the estimate. Get help with specs if you're not comfortable doing them yourself.

You should expect to get your estimate from the printers in 24-48 hours—they can do it for most jobs, really they can! When you send your specs (email, fax, or mail them), ask the printer if he or she has enough information to quote the job.

The estimates you get back will never be exactly the same, but they should be close. If one sticks out as very high or very low, I'd be leery of giving that printer my job. The "low bidder" may be "low balling" you to get the work. It's not always true, but usually you get what you pay for.

Should you get estimates for every print job? That depends. If you've established a good relationship with a printer and you trust him or her to deliver high quality on time, then no, I wouldn't get estimates for every job. But I would get them now and then. You need to be sure the pricing is still fair.

So get good estimates by giving good specs to a few printers. Many printers now have spec forms for you to fill out via their web site, which makes the process a whole lot easier.

P.S. Your favorite printer may not be the best source for all jobs. What if you normally print flyers and letterhead, and suddenly you're in charge of the annual report? Check around for printers who do a lot of annuals.

Back to Basics for Buyers:
Writing Specs

If you want an accurate estimate from a printer, then give him or her a lot of details about the job you need produced. These details are known in the trade as "specs," which is short for specifications.

Most printers have their own spec forms. Some are more detailed than others. I've seen forms with questions that only printers or very experienced buyers and designers would know how to answer.

A printer made the suggestion that buyers ought to have a specifications form (spec form) handy when they're asking for estimates. Great idea!

Phil Green covers this topic brilliantly in his book, *Quality Control for Print Buyers*. Below, I've summarized his recommendations for a complete spec form. If you'll be buying print regularly, why not print up a batch of spec forms and keep them handy? Not all of these specs are applicable for every job you create, but you have to start somewhere.

Job Name—Pretty obvious. Name your job for easy referral.

Due Date—Specify when you need delivery, in case this date cannot be met. How critical is this date? Is the job for a specific event? Don't forget to figure in mailing dates.

Quantity—It's smart to ask for prices on a few quantities from the get-go, rather than just one. (It saves time for both of you.) If you think you want 2500 pieces, ask for estimates on 2000 and 3000 while you're at it. You might decide to print more if the price is right.

Finished Size—Cite the dimensions of your finished piece. Be sure to note whether these dimensions are flat or folded. For example, a 4-page newsletter is probably 11" x 17," folding to 8 1/2" x 11." It could fold in half again to 5 1/2" x 81/2," or you may want a letter fold. Be specific.

Page Count—If known. Remember to think in multiples of 4 pages. For books/booklets, make sure you specify text pages + cover.

Number of Inks—Will it print in black ink only? Or is it 2 colors? Maybe 4-color? Does your company have a "corporate color" or some special ink? If you're printing a booklet, will you want more colors on the cover? Not sure? Ask your printer early on. Will any of the inks "bleed" off the paper's edge? That can cost more, too.

Images to Be Scanned—How many photos or other images does this job include? Where exactly are they coming from?

Proofs—What kind of proof(s) do you need to see?

Here's some "decoding" that might come in handy: if a printer writes "1/1" it means 1 color on both sides of the sheet; "2/2" means 2 colors on both sides; "4/4" means 4 colors on both sides (this usually means 4 process colors). "1/0"means 1 color on one side and nothing printed on the back side.

Paper—Get as specific as you can about paper, with your printer's help. Paper can account for 30% to 60% of a job's cost. Just specifying "cover" or "text" weight is not enough. If your job needs more than one type of paper, let the printer know! Paper comes in different grades, weights, brands, finishes, and colors. If you can't specify exactly what you need, ask for help.

File Format—Are you creating the job on a Mac or a PC? In what software? What version of this software? Do you want the printer to typeset/design the job? Or are you planning to give a printer hard copy (camera-ready copy) instead? Do you know what he/she needs? Discuss this early on.

Binding—Will your job need to be bound? What kind of binding, exactly?

Finishing—What kind of finishing will your job need? Does it need to be folded? Does is need laminating? Varnishing? Die-cutting? Embossing or engraving? Three-hole punching?

Shipping Requirements—Does the job need to be shrink-wrapped? Shipped to several places? Boxed in any special way?

Other Comments—What else should the printer know about this job? Are you doing a series of similar jobs every month or so? If so, will you need more of the same paper for these jobs? Here's where you note any special requirements you may have.

A print job is the sum of many parts. These parts are the specs, so give your printer as much detail as you can when planning a job. Remember, every print job is custom built. Next time you're tempted to ask a printer to price a job, think of all those details that matter.

Why not create your own spec form and ask three printers to supply estimates in writing? Then you'll be comparing apples to apples. When specs change, inform your printer and ask for an adjusted quote.

Forget the Price Break—Print What You Need!

"When does the price break kick in?" is a question I am often asked by folks looking to buy printing.

"How many do you need?" I reply.

Sure, the more you print, the lower the unit cost. But if you don't use all 1000 flyers, or if your zip code changes long before your five reams of letterhead run out, who cares how low your unit cost was? You've wasted money!

One client regularly printed 10,000 of everything—simply because the price was "so good." He had no idea if he'd ever use all of these materials. No kidding.

Take the time up front to determine a reasonable quantity of printed materials. Figuring out a three- or six-month supply is a good idea. And ask your printer about reprint charges.

Plus, the more often you print, the more up-to-date your materials can be.

"Overs" the River and Through the Woods

The printing industry has used standard trade customs and business practices for more than seventy years, although few print buyers are aware of these customs. There's one rule about quantities that's worth knowing.

The rule states that over-runs and under-runs ("overs"

and "unders") will not exceed 10% of the quantity ordered. This means that a printer is within his or her rights to deliver as much as 10% more—or less—than you ordered.

It's also customary for the printer to bill you for the amount that he or she delivers within this range.

Just be aware of this rule when ordering quantities. If you need exactly 1,200 brochures for a mailing, for example, you should order more than that—or run the risk of coming up short.

How Shall I Say This . . .
Size Matters!

Fact #1: While producing shareholder reports for a previous employer, I asked my printers to tell me the optimum finished size for their particular presses. By altering the format, I helped save my employer hundreds of thousands of dollars annually.

Fact #2: *The Wall Street Journal* reported that some major daily newspapers are shrinking. To save money. Big money. This national newspaper trend to reduce page widths— some by as little as a single inch—will save publishers millions.

Why should you care? Because this "size lesson" is mighty powerful. By changing your print job's finished size by a fraction of an inch, you could conceivably save thousands, hundreds of thousands, or millions of dollars.

Since the printing industry revolves around the 8 1/2" x 11" size, multiples of this size are the most cost-effective. As a rule of thumb, the odder the finished size, the larger the price tag.

It never hurts to ask your printer: is this size the most cost-effective for my job, or do you recommend another?

Chapter 5

WORKING TOGETHER

The more you understand the printing process, the better your experience with printers will be. This chapter covers such important topics as proofs, press OKs—and thanking your printer for a job well done.

8 Ways to Work Better with Printers

This really happened. A client once screamed for 20 minutes about his bad experience with a local printer. "They're medieval!" he ranted. "Printing is like alchemy—the printer disappears with your job behind the pressroom doors, and you've no idea what's going on back there! They communicate nothing." He'd decided the whole industry was evil.

He had reason to complain, for the printer had delivered a shoddy product. But I hated the way he generalized about all printers.

He could have had a successful experience if he'd cooperated with his printer in a methodical way. Here are 8 tips on how to do just that:

1. **Good printers are the rule, not the exception.** Most

printers work very hard to deliver what customers want. They have an amazing amount of technical expertise. If you haven't found a great printer, keep looking.

2. **Make it personal.** Find a salesperson you like. He or she should be experienced, reliable, honest and reachable—you don't need someone who seems to go AWOL on you. And pick one who'll go the extra mile.

3. **Keep your printer in the loop.** One of the biggest mistakes that print buyers make is failure to involve the printer soon enough. The success of your print jobs largely depends on your communicating early and often with your printer.

4. **Play fair.** Since every job is a custom job, respect the fact that it can take time to print something well. Don't cry wolf and impose artificial deadlines, when in reality you could wait another day or two.

5. **Be specific.** Every detail about a print job affects its price: the format, number of pages, quantity, inks, paper, folds, etc. Put someone who's detail-conscious in charge of your printing. And give the printer all of the specs up front. Don't eke them out over days or weeks.

6. **Discuss file formats.** How you prepare your files is very important to a printer. Are you using a Mac or a PC? What software, specifically? By having this discussion early on, you can make sure that what you're creating is printable.

7. **Ask questions.** Don't be intimidated by the language. Unless you're used to dealing with printers, chances are you'll need lots of terms translated. If you don't understand something, ask!

8. **Be clear about responsibilities.** Clarify what your role is vs. the printer's. A businessman I know had

his printer do some typesetting. He approved a proof of the job, but he failed to notice a typo that the printer had typeset. He OK'd the job to print. When the job was delivered, he noticed the typo (but of course). The printer wasn't responsible; the client was. Proofreading is almost always a client's responsibility.

How Customers Can Cut Print Production Time

I often hear complaints from customers who wonder why print jobs take so darned long. Once I do a little digging, it's almost always clear that they share the blame for much of the delay.

Here are six basic rules of thumb for working with your printer to save production time:

1. **Be specific with your printer from day one.** Tell him or her what you need done, when you absolutely need delivery (be honest, don't make the printer jump through hoops for nothing), and how you plan to deliver the job. What software will you use to create it? Your printer needs to know; it will definitely affect production time and costs.

2. **Tell your printer how you're going to use this piece.** Once it's printed, does it need binding or other finishing? Drilling? Die-cutting? Does it need to be addressed and mailed? What is its final destination? Will it be inserted into a mailing package with other material? Printers have connections, capabilities, and lots of resources. Many can handle mailing and fulfillment on site. Give your printer the whole story early on, so that you don't waste time on the back end of the job. For example, you may be sending your job to a mail house for addressing and mailing,

when your printer could address the pieces as they come off the press.

3. **Have good job specs handy when requesting price estimates.** If you're not sure of the quantity, ask your printer to quote on a few quantities (not a huge laundry list, please!) so that time isn't wasted by your requesting new quantities day after day.

4. **Ask for a schedule.** Stick to it! It needn't be formal, depending on how simple or complex your job is. Work backward from the due date. The schedule should include key dates, like delivery of your file, proofs due from the printer, proofs approved by you, and final delivery date. You may have a delivery date in mind for a good reason: an event, a mailing, or you've run out of a printed piece and need more ASAP. If so, tell your printer that the date is critical, and that if he or she cannot promise delivery, not to take the job.

   Make sure you give your printer enough time to do the job. Very few print jobs can be produced in a few days. I recommend asking a printer, "Is this delivery date reasonable?" If the answer is yes, then you should expect delivery as promised.

5. **Turn your proofs around ASAP.** This usually means the same day you get proofs from your printer. If you hold up your proofs, the printer may not be able to make your delivery date. (Try to limit the people on your end who have sign-off responsibility, too. The more people who look at a proof, the more likely someone will want to make changes.)

6. **Keep AA's to a minimum.** AA's are Author's Alterations, which are changes you make to a job after it's been sent to a printer. (Expect to pay for AA's, although not every printer will charge you.) Save time by preparing your copy carefully before

you send the job to a printer. Lots of people wait until they see their copy typeset before editing it: inefficient idea! This will seriously eat up your production time and will cost you more, as well.

Problems do happen in printing, and dates get missed. But by having an open dialogue about your expectations with the printer, you'll avoid being surprised by something like a missed delivery.

In a perfect world, your sales or service rep would keep you informed, but they're juggling lots of jobs every day. Check in with them a day or two before the delivery date.

If you give your printer enough time, turn your proofs around quickly, don't make significant changes, and there are no press problems, yet your dates are missed, it may be time to look for another printer.

A Primer on Proofs

When you give a printer a job, ask to see a proof of that job before it actually gets printed. A proof is your chance to carefully check for color, content, images, and layout before a job goes on press. You may see one proof or several. Ask your printer what you're expected to do with a proof and how closely it will resemble the printed piece. Typically, a customer makes marks on a proof to indicate mistakes or last-minute changes. Once you approve it, sign and date the proof and return it promptly to the printer. If you hold it up, you may jeopardize your schedule.

There are several kinds of proofs available today, and different printers offer different choices.

Most proofs are still hard-copy proofs—ones you can hold in your hands. These are either analog (photographic) or digital. Printers generally deliver these to you via courier or the mail.

There are soft proofs, too—those you look at on your own computer.

The newest kind of proof is the remote proof, whereby a printer transmits your job to you digitally, and you then print it off via a proofing device in your office. Color management and high-speed delivery are key issues in remote proofing. An excellent, in-depth article on this topic was written by Allen Avery in the February 2002 issue of *American Printer* (www.americanprinter.com).

The following lesson in "Proofing 101" is based on information I got from Jamie Bradley, Print Management Director for Fusion in North Reading, Massachusetts (www.fusionpromo.com).

The three main categories of proofs are photographic, digital, and Web-based.

1. **Photographic (or Analog) Proofs**

    If your printer is still working with film rather than direct-to-plate or computer-to-plate technology, you'll likely see analog proofs. They date from before desktop publishing, when artwork was pasted onto matte board and photographed on a large-format camera to create the film negatives used to make printing plates. In general, these proofs are labor-intensive to produce, because the film needs to be stripped into place, exposed one color at a time under special lights, and then prepped for final presentation to the end user.

    *   BLUELINE (also known as a blue, Dylux, saltprint, silverprint): Blues are the familiar yellow paper proofs with faded blue type. They can be produced in large sheets and folded or cut down to show the final size and page layout of your piece.

- VELOX (also known as white print): These contact prints are used to show accurate dot resolution in halftones, usually shown in conjunction with a blueline.

- COLOR KEY (also known as a 3M or color overlay proof): Consists of layers of loose plastic film taped together in registration (so that the colors are positioned correctly together) for both spot and process colors. They're especially helpful when you want to check exact color breaks (that is, when you want to see where the separate colors appear in your materials).

Matchprint, Chromalin, Waterproof, Chromacheck: These are the trademark names from different companies for laminated color proofs. Process (or spot) colors are exposed individually onto film transfer sheets, laminated in register to each other and to a paper backing sheet. They are the most accurate representations of the final printed piece. The specific proof you receive is based on the system owned by your printer.

2. **Digital Proofs**

   Even printers who still use film to make their plates are making the switch to digital proofs because of lower costs and faster turnaround. The technology has evolved substantially since the earlier days of desktop publishing and access to excellent equipment is within the reach of nearly every printer.

- LASER PROOFS (Xerox, Phaser, Canon): These are high-quality color copies that can be reasonably representative of the final printed piece, depending on how hard the printer works to calibrate between how they proof and how they print. Inexpensive and quick, laser proofs are gaining wider acceptance as a proofing tool as the manufacturers work to tweak their systems so that the bright pigments they put in their toners for the consumer market are adjusted to represent what can actually be printed in process colors.

- INKJET (Epson, HP, digital Dylux): These proofs have taken the place of the blueline, since they can be printed on both sides and folded or cut down to show the final size and page layout of your piece. Don't expect accurate color.

Dye Sublimation (Rainbow, Fuji, Approval, Matchprint, Chromalin, Waterproof, Chromacheck): Dye-sub proofs use blended CMYK waxes/resins in conjunction with heat to produce either continuous tone or dot images, giving the best representation of the final printed piece.

Thermal Wax (Duoproof, Polaproof): These proofs use a waxy material and can print on a wide variety of media. They're less expensive than dye-sub proofs but are more limited in their ability to accurately demonstrate how colors will reproduce on press.

3. **Web-Based Proofs**

As more print-on-demand and variable data applications are moving to Web-ordering systems, it has become necessary to have accurate proofing methods that don't change formatting or line breaks when files are processed. The most common formats for proofs are PDF, TIF, and JPG, with some companies offering proprietary formats. Don't trust these proofs for accurate color, but use them to check copy and layout.

There are two other proofing terms worth mentioning.

Contract proofs can be produced by any of the above-mentioned methods, but the term generally refers to the proof agreed to by the client and printer that will be most representative of the finished printed piece.

Press proofs are really not proofs at all, but finished printed pieces that are printed on a press. They're used for color checking at short-run costs before reprinting a longer production run.

No matter what type of proof you get, ask your printer's advice on what to look for, how to mark it up, how expensive it will be to make changes, and how closely it will resemble the final product. You want to avoid surprises (and disappointments), right? Believe me, so does your printer.

Avoiding the Blues with a Blueline

Q: If a printer sets type for your job, misspells a word, shows you a blueline proof, and you OK the proof—who's responsible for that mistake on the printed piece?
A: You are.

The purpose of the blueline (aka blueprint, silverprint, or Dylux) is to allow the client to double-check all of the content of a job to be printed to make sure it is correct. Even if the printer set the type.

Here are key things you should check on a blueline:

— all of the text
— headline placement

— photo placement/cropping
— all other images in your piece
— sequence of pages

Sometimes a blueline is the last proof you'll see, so take your time and check every element carefully. Ask your printer how long you'll have to study the blueline (often he'll want it back within 24 hours), and then make it a priority.

Check all folios (page numbers), all levels of headlines, and all the display type (such as headlines) for alignment. If it's a multicolor job, check the type that will print in other colors (this will be obvious).

Check photos to be sure each image is correct and cropped properly, and check captions, too.

Mark every correction clearly on the blueline. Circle blemishes and broken type. Be sure to sign your name and date it, too.

If you're not sure what to check on a blue, ask your printer. Once you sign off, it could be an "OK to print." If you've missed errors on a blue, you are usually liable.

Finally, a blueline isn't an opportunity to rewrite your content—that will cost you dearly! It's the time to make certain that all of your copy and graphic images are where they should be.

AA's vs. PE's—Vive la Différence!

So there you are, standing in the print shop, ready to hand over a job to be printed. It could be hard copy. It could be a digital file that you've created on your PC or your Mac. You're glad to be getting this job off of your desk, because you're behind on it, and you'd just as soon proofread it when the printer sends you a proof. Besides, isn't it easier to read once it's in that nice format? Then you can really see what your job will look like.

In a few days, the printer sends you a proof, maybe a blueline (or Dylux), and asks you to review it carefully, because the next time you see this job, it will be printed and delivered. You take this proof and proceed to pass it through the department. Maybe now is the time when you decide to get your boss's input on it.

Every Tom, Dick, and Harry in the company gives his two cents to this proof, and by the time you get it back, it has so many editorial comments on it that it's hard to distinguish these last-minute changes from the first draft of your copy.

Back it goes, changes and all, to the printer. You ask to see another proof. It arrives in a day or so. You bless it (with just a few "minor" corrections), and finally, it's ready to be printed.

The printer delivers the job. You love it. Your boss loves it. Then the invoice shows up, and you spot the charge for AA's.

Before you pick up that phone to call your printer with a "What's the meaning of this?" tone in your voice, know this:

Changes that you make to a job once you've already given it to a printer can rightfully be charged back to you. Such changes are known as AA's, which stands for Author's Alterations.

On the other hand, if corrections are needed on a job after the printer has it and are attributable to the printer, those will not be charged to you. They are known as PE's, or Printer's Errors.

Not every printer will charge for AA's, but be prepared to pay for them. Proofread everything more than once before you give a job to a printer. Have someone else proofread it . . . every word, every number, every bit of punctuation. If you keyboard it, you own it! Printers are not responsible for proofreading. You are. If you've made mistakes and want to correct them, expect to pay for them.

Doing a Press OK:
Advice for New Print Buyers

For most print jobs, you're not in the plant when the job prints. But if you have a 4-color job, or one that may be difficult to print, or a job that requires premium quality, you might want to be at the printer's doing a press OK, which is also called a press check. Some people press check 2-color jobs. It all depends on the job.

You should be accompanied by a sales or service rep when you do a press check. He or she will guide you through the process, which can take 15 minutes or last for several hours.

Here are a few ideas about how to conduct a press check.

Let your printer know early on if you plan to press check a job. The cost will be built into your job. But always be efficient when on press: future job estimates might be higher if you tend to waste time on press checks. Be friendly, but remember that it's business. Everyone wants to get your "OK to print" quickly so that the job can run.

By the time you do a press check, you'll have signed off on the proofs that the printer provided. Compare these color proofs with the press sheets you'll be shown. Buyer Jill Connolly (Analog Devices) suggests that you ask the printer how closely the proof is expected to match the printed sheet.

Walter Burroughs, a printing consultant who conducts seminars on doing press approvals for customers of S&S Graphics in Laurel, Maryland (www.snsgraphics.com), has a lot of solid advice for new buyers.

Make sure that the paper being used is the paper you specified, Burroughs says. He even recommends that you inspect the paper labels from the cartons.

Proper lighting is critical. You want to view the sheets under 5000 Kelvin lighting.

Speaking of color, here's a handy tip from Jack Fleming of Pomco Graphic Arts in Philadelphia, Pennsylvania: don't wear a color like red or bright blue on press. The paper will actually reflect reds, blues, and greens. The best colors to wear on press are neutral colors. "Did you ever notice," Fleming wrote, "that pressmen wear grey, light blue, or other neutral colors on press?"

Now is not the time to check for the correctness of your content. You should have done that during the proofing stage. You do a press OK to approve the color and make sure that the job is printing as planned.

"You mark everything wrong on the press sheet and start the inspection by writing the number 1 on the first sheet," said Burroughs. Why? Because you'll be seeing several press sheets and they'll all look very similar. By numbering them, you'll keep them in chronological order and be able to track the changes made on press.

Circle all extraneous marks and hickies (they look like tiny white donuts and are caused by dust or bits of paper).

Buyer Karen Casale McLaughlin of Talbots reminds buyers to pay close attention to this first sheet. "Be aware of the color bars and how they are "reading" prior to making color adjustments. Get oriented to which way the sheet is printing . . . number your sheets as you proceed, so you know where you are and where you've been." McLaughlin adds, "Take your time. Make sure the color's balanced from one side of the sheet to the other. Check register, layout, images and type (not for correctness but for ink density and overall appearance). Remember, too, that when you make a color adjustment, you in turn affect the color in that inking row."

Have your sales rep with you when you're "press-side." He or she should be able to help you interpret what you're looking at. You may also be working with the pressperson as well. Some print shops have customers do press OKs in special viewing rooms, while others bring customers right into the press room.

How do you communicate changes to a pressperson? All of the people I spoke with agreed that speaking in plain English is best. No matter how familiar you are with CMYK (the four colors in process printing), don't try to "outprint" a printer. By that I mean: say something looks "too red" or "too dark." Let the pressperson make adjustments to the CMYK. Use your own words to describe what looks "off" to you, and trust the pressperson to fix it.

"Presspeople know their presses better than anyone else," Connolly wrote. So trust them. At the same time she cautions buyers not to "be bullied." If you're not happy with the press sheets, speak up. Let your sales rep help.

Connolly suggests that you check trim lines, too (lines that indicate where the pages will be cut or trimmed), and ask for a rule-up of the piece so you can see that everything fits as it should. Ask for a quick dummy, she adds, to check the imposition (that is, so you can see how the copy and the pages will be arranged in the finished piece).

Once you're satisfied with a press sheet, sign off on it with an "OK to print" and date it. Take a few of these sheets with you and check them against printed samples when they arrive.

Is this an exact science? Hardly. And different buyers have different advice and idiosyncrasies about conducting press checks. Just remember to be reasonable while on press—and as Karen McLaughlin points out, "Say thank you to whoever worked with you on press."

The more experience you have, the easier it will all become. Ask lots of questions and learn from the people running the presses.

Fear of Press OKs

Thinking back, I remember how scary it was for me, as a new print buyer, to do a night-time press OK. Here's how it went:

I was ushered into a darkly lit waiting room by someone other than my salesperson and told it would be a few minutes until I got a sheet to review. Eventually, two heavy black doors swung open and a pressman brought me a sheet to OK. He stood back, not saying much, and I began the slow task of scrutinizing every image, every bit of content, every color, on every page. Squinting, reading, double-checking the sheet against my OK'd blueline and/or previous proof, I hoped I wouldn't see anything that made my throat catch.

When I questioned a full-color image that didn't seem quite right (I never pretended to speak the secret language of printing), the pressman (never a presswoman, unfortunately) said he could "bring up" or "bring down" this or that color, and out he'd go for another 30-50 minutes, until we started the process again.

A handful of sheets and several hours later, I'd say "OK" and sign a sheet, and we'd both heave a sigh of relief.

Nine times out of ten, everything worked out well. Occasionally, I'd have to compromise an image, color-wise, which left me to explain to my bosses (in layman's terms) why the corporate color wasn't quite what they'd envisioned, or why we saw a bit of ghosting on a solid. (Ghosting is when a faint or "ghostly" image appears on a printed sheet in a place it's not meant to appear.)

Press OKs were hairy. Part of it was because I didn't have a technical background. Not knowing what was causing the colors to shift ever-so-slightly with each new press sheet was intimidating and frustrating. Part of it was realizing what a heavy responsibility it was, for most of my OKs were on expensive annual reports in quantities over 200,000.

What would have helped? I would have liked the following: a bit more warning about what to expect, more knowledge about what was and was not reasonable to ask of the pressman (and the press!), better lighting with which to do a press check, and the company of my salesperson.

I was lucky to have worked with some mighty fine pressmen. In particular, I remember Mr. Ed Garneau of Nimrod Press in Boston. Ed had a way of making the whole experience OK in every sense of the word. He never made me feel stupid. That was worth a lot.

Even Exact Reprints Can Go Awry

Don't assume that ordering an exact reprint is always problem-free. "Things happen," as one print customer noted.

He ordered an exact reprint of a full-color jumbo postcard that included a fifth color, a PMS gray covering the whole background of the card with the process color in the middle. It was done by the same printer, just one year later. The customer asked for a blueline (proof), but was told that it was company policy not to show one for exact reprints, and that the reprint would match the original.

But the color didn't match. The PMS gray was "colder and bluer" than the first printing.

Nine times out of ten, straight reprints will go without a hitch. But this real-life example is a reminder of a few key things. First, always see a proof or two. A blueline wouldn't have prevented this particular problem, but color proofs might have. Certainly, a drawdown would have made a difference. A drawdown is a sample of a specific ink color, printed on the exact paper for the job. It's shown to a customer for his or her approval before the job's printed.

Seeing the job on press would have helped, too. Noticing the color problem there, the customer could have worked out a solution with the pressman. It turns out the sales rep did the press OK on his own, and although he noticed the color issue, he ran the job because they were indeed matching the color spec'd (PMS 427).

In the end, the gray didn't match, said the printer, because the varnish that was mixed in with the PMS 427 yellowed over time.

There's definitely a lesson or two here. First, time changes things in printing. Ink colors can fade. Even the exact same paper, I'm told, can change from one year to the next, like "identical" lots of yarn coming from the same manufacturer.

Second, certain PMS colors are more likely to change. Jack Fleming of Pomco Graphic Arts in Philadelphia, Pennsylvania, noted that "PMS 427 is about 98.4% transparent white. Any color that has that much white in it is guaranteed to change. That change is usually yellowing." And don't forget that different editions of the Pantone Color Formula Guide will show slightly different shades of the same color, because those printed editions fade, too. So the edition that you're using might look quite different from your printer's edition!

Here's the bottom line: things can go wrong with exact reprints. Both you and your printer need to be proactive to prevent possible problems such as this one. Insist on proofs or drawdowns if warranted, and see press sheets if you can. This printer should have warned his client that the colors he expected probably wouldn't be an exact match.

Checking Your Printed Product:
What to Look For

If you're new to print buying, here's a handy list of what to look for when you receive newly printed materials, to make sure that the print quality is good.

These 10 tips come from David Nicholass, former Color Quality Manager (best say "colour" manager!) at the *Financial Times* in London:

1. Check the overall look of the product. Is it as you expected it to be, or are there areas with which you aren't totally satisfied? If so, use this list to help you pinpoint any faults.
2. Check the colors. Are they even? Does the final piece match your proofs?
3. Check the registration if multiple colors are used. This means that all colors are printed on top of one another. Use a magnifying glass to see if any colors are "peeking out" from under other colors (which would be an example of bad registration).
4. Check the binding if you have had a book or magazine printed. If the book is "stitched" (stapled), are the wires sitting snugly in the spine of the product, or are they "creeping" onto the page? If your product is perfect bound (flat, glued spine-think "paperback book"), can you open it without having to rip it at the spine?
5. Now the trimming. Most printed products have to be trimmed (cut to size) at some point. If you have a leaflet or sheet produced, is the trimming even and square? Have any of your elements literally been cut off? If you had color printed on the background, is there any white showing at the edges? All these point to bad trimming.
6. Folding. Inaccurate folding can ruin the appearance of a product. When you hold your product as "finished" (closed magazine, folded letter for instance), is it folded neatly? A magazine folded correctly will have neat, lined up edges for all of its pages. Pages should never "fan out."

7. Over-inking. Evidence of over-inking can be heavier, darker images or "offset." Offset occurs when so much ink has been applied, it doesn't have time to dry, so it transfers onto the opposite page or, in the case of a stack of leaflets, onto the back of the leaflet stacked on top of it.

8. Is everything where it should be? Silly question? You'd be surprised. Pages can be inserted upside down; you could have two identical pages; and a whole host of other faux pas. Check your product inside and out.

9. Take random samples. Remember, you'll be sent "file copies" or specials from your printer. If you think you have cause for complaint, ask to see "timed copies." This will give you an indication of the entire run's quality.

10. Take "commercial acceptability" into consideration. Printing is a process involving ink, water, paper, and lots of moving parts. Things can and do go wrong, and this is sometimes beyond your printer's control. Don't be any fussier than you need to be.

The above points are general rules of thumb. You have to take into account the materials used and the processes involved.

Remember, if it looks good to your eyes, it will look good to your intended audience as well. Work with your printer, not against him or her. If you spot a problem on your samples, call your printer and discuss it as calmly as you can.

Have a Heart:
Pay Your Printer Promptly

Let's be frank—unlike cheese, an invoice does not "improve" over time. Yet some print customers sit on their printing bills for months. Want to be a good customer? Pay your bills promptly. It could improve your pricing.

How so? Here's what Jack Fleming of Pomco Graphic Arts in Philadelphia, Pennsylvania, told me: "If more customers paid in the 30 day net terms that appear on most printers' invoices," Fleming wrote, "we could offer better pricing."

Printers need the money that customers owe them to operate their businesses. Without it, they have to borrow cash and pay it back with interest. That interest then has to be factored into their bottom line.

Fleming suggests that companies offer discounts to encourage early payment. If customers pay within 10 days, they get a 1% or even a 2% discount. This translates to substantial savings to buyers over the course of a year.

"I can't tell you how many customers don't even look at the invoice until they receive a statement after 30 days, reminding them of their obligation," Fleming notes. "And then they want to contest AA's, which delays payment even further."

This highlights another problem that occurs when customers sit on an invoice: the job's so old that it takes a lot of backtracking to reconstruct what happened during the production process. "We have to go back, check the paperwork, and get the answers. You can imagine that this takes time, and the invoice gets older. Sometimes it's tough to pull out all the facts on a $239 AA after 30 or 40 days. It's especially disconcerting when that $239 AA holds up a $25,000 invoice."

One printing company has a new procedure for processing invoices promptly. Before they invoice the customer, they email or fax a copy of the bill and then follow up with a phone call to see if there are any questions. It's easier to answer such questions while they're fresh in both parties' minds. Then they mail out the invoice, and on or about the 25th day, they follow up with a friendly call to check on the status.

Fleming encourages customers to be up front with their printers about when they expect to pay their invoices—even if they plan to pay in 60 days. This lets printers plan ahead.

So don't keep shuffling those printer invoices to the bottom of the pile, especially if you might have questions. Deal with them promptly. Perhaps you'll be offered an incentive in the form of a discount. You'll certainly be considered a good customer, which usually translates into more TLC from your printer.

Say Thanks

A seasoned print sales executive from Boston offered this valuable insight:

"A simple thank you note can work wonders. Printers post such notes for all their craftspeople to read. And when individuals are called out for special recognition, such notes are particularly welcome. Plus, the writer (print customer) wins a special spot in the printer's heart, as few accounts take the time to acknowledge a job well done."

How easy to do! Buyers, take the time to acknowledge a job well done by your printer. A hand-written note, an email cc:ed to the company president—even a one-minute call to your salesperson—can make a big difference. Don't let bad news be the only news that travels fast. Did your CSR find a typo on a blueline? Did your salesperson recommend a printing alternative that saved you big bucks? Did a prepress wizard undo that mess of a file you sent over? Acknowledge them all!

They'll remember you forever.

Chapter 6

PAPER AND INK

No matter how a printer gets there, he or she creates products by putting ink on paper. But don't be fooled into thinking that it's child's play. First, the printer has to plan how the job will print, price it out, schedule it internally, select the right paper, communicate all of this to you, the customer, and then production can begin. Now it's "ink on paper" time. But it doesn't happen by chance.

When you realize how all of the steps are interrelated, you begin to see what an incredible mixture of art, science, and technology printing really is.

Paper, Paper! Read All About It!

When choosing paper for your print materials, there's a lot more to it than coated vs. uncoated or color vs. white.

Here are eight basic rules of thumb to keep in mind:

1. Paper typically constitutes 30% or more of your bill, so choose very carefully.
2. Using a printer's "house sheet" can save you money,

but since it may change, be sure to inquire about it every time.

3. The weight of your paper can impact your mailing costs.

4. Avoid using coated paper for such things as forms or reply cards. Ball point pens may not work; flair pens will smear.

5. Make sure your paper is laser compatible if you plan to use it in your laser printer.

6. The industry revolves around the 8 1/2" x 11" sheet of paper, for the most part, so the most cost-effective jobs are multiples of this size.

7. Uncoated is cheaper than coated paper; white paper is cheaper than colored paper; and lighter-weight paper is cheaper than heavier paper.

8. Paper has the ability to alter your printed image. Beware!

To be safe, always consult with your designer and/or printer when choosing paper.

Figuring Paper Weight Is a Heavy Duty

Making sense out of paper weights is confusing for new print customers. Here are a couple of helpful guidelines:

Every type, or grade, of paper—such as bond, cover, and text, to name a few—has one basic sheet size that's used to figure out its basis weight. Basis weight is the weight of a ream of paper (500 sheets) in a grade's most standard size. In the US, it's expressed in pounds. When written down in specs for a print job, the # symbol stands for "pounds."

Let's say your stationery is 24# bond. This means that a ream of this 17" x 22" paper (its basic size) weighs 24 pounds. Many grades are made in different weights, too. Common weights for bond are 20# and 24#, but several other weights are also available.

Sometimes you may want to substitute another type of paper for a job you're printing. Maybe you're in a hurry, and your printer can't get a specific sheet of paper in time. Or you simply want a different look and feel for your job, and you want to substitute another kind of paper.

You can do it! There are equivalent basis weights between different paper grades. They're equivalent because, in general, they're about the same thickness and in that way are interchangeable. Your printer or designer can advise you on which paper you can substitute. For example, 50# book paper is equivalent to 20# bond paper. (Get a sample of each, close your eyes, and study them with your hands.) The ream weights are different because the standard sizes are different.

You have lots of options when it comes to paper weights. In general, heavier paper costs more. Mailing costs should be considered, too. You should to get a dummy weighed at the post office before ordering a large print job that's to be mailed. Using a lighter-weight paper can save you significant postage.

Paper choice matters a lot. It pays to know your options.

Stock Tip:

Making Sure Your Postcards Pass Muster

A client emailed me her pet peeve: "Which card stock is the right weight for a postcard?"

Little did she know how hard it would be to give her a straight answer—but here goes.

Postcards and reply cards must measure at least 7 pt. in thickness, or bulk, according to the U.S. Postal Service. Paper thickness—called caliper—is measured in thousandths of an inch and expressed in points, with one point being .0001 inch. So postcard stock must measure at least 7 pt. (or .007 inches).

Yet when you buy or specify paper, you often do so by the paper's weight, among other characteristics, not by its bulk. For example, you might print your newsletter on 60# offset and your pocket folders on 80# coated cover.

But a paper's basis weight, such as 60# or 80#, has nothing to do with its bulk. So the onus is on the consumer to make sure that postcard stock measures up to the post office's standards—that is, at least 7 pt. for standard postcard size.

There's no guarantee that every 65# cover or even 80# cover stock will qualify, since thicknesses vary among paper types.

What to do? Check with your printer or with the post office before you start printing those cards.

P.S. The USPS publishes a variety of helpful guides on different topics, such as *Designing Reply Mail* and *Designing Letter Mail*. These guides are free and should be available at your local P.O. or at the web site www.usps.gov.

Dot's the Truth

One term that you'll often hear, if you buy printing professionally, is "dot gain." I think it's unique to the printing industry.

In offset printing, a photograph is converted to dots. During the prepress and on-press stages, these dots tend to get bigger, or spread.

This phenomenon is known as dot gain.

In order to give you the printed quality you anticipate with photographs and other continuous tone illustrations, a printer has to compensate for dot gain by making the dots smaller in the digital files and also in the film before the job prints.

Quality can suffer in several ways if a printer doesn't adjust for dot gain—mainly, your pictures will lose their detail, and printed colors won't be what you expect.

Adjusting for dot gain is complicated because it is affected by the paper you print on, the type of printing press that's used, the printing process used, and the inks themselves.

Want to impress your printer? Ask for his or her dot gain "prediction" when you're printing photos.

Dot's it for now . . .

"... That's Amore!"

Wasn't it Dean Martin who sang, "When the moon hits your eye like a big pizza pie—that's amore!"?

He probably wasn't singing about a printing problem known as a moiré, which is an undesirable screen pattern that occurs when halftones and screen tints are made with incorrect screen angles.

A moiré pattern may also develop if a printer screens a photo that was already printed in, and therefore screened for, another publication. It could also happen if a pattern in a photo—like a plaid—interferes with a halftone or dot pattern.

A moiré screen often looks like a blurry plaid pattern, or a "dotty" pattern, and it needs to be corrected before your job prints.

So if your printer starts "singing" about "a moiré," he's not making a pass at you—he's just alerting you to a problem.

Please Don't Be Rubbed the Wrong Way!

"Print rub" is a technical problem that's caused by solid ink coverage or 4-color process printing on matte or dull-coated stock. As the name implies, the ink tends to rub off these surfaces after it dries.

The problem lies with how these papers are created. Jim

Hamilton of Quebecor World Universal Press explained it to me this way:

After the matte or dull coatings are applied, the paper surface is roughed up to dull down the gloss. Whereas a gloss coating appears smooth under a photomicroscope, matte and dull coatings are craggy, like hills and mountains.

After the ink dries, the solvents evaporate, but the ink pigments sit on top of the paper and are prone to scuff or rub off the tops of these hills and mountains.

What's the solution? According to Jim, spot varnishing should be mandatory for any 4-color process, solid ink coverage, or areas that bleed off the trim on matte- or dull-coated stock. The varnish protects the surface and seals in the ink. (See "Varnish: More Than a Garnish" in Chapter 9.)

If a job is printed on a press where inline gloss varnish is not available, be advised that marking or rubbing is inevitable to some degree.

Note that cast-coated stocks should always be varnished for the same reason. And be careful when using a lot of reflex blue, because more pigment is left on top of the sheet. It's better to replace reflex blue with an imitation blue with less of a potential rub problem.

So avoid printing rush jobs on matte or dull-coated stock, and always seal in the ink on these papers with varnish.

Why Does Paper Crack?

You've spent months producing a high-end marketing piece. Everything goes smoothly on press. The color's perfect. The samples arrive early. And then you spot it: that long, white crack all along the fold. The fibers of the paper are showing through.

What happened?

To help explain why paper cracks, I spoke with Denis Doutre, Director of Field Technical Services for Domtar, Inc.

Domtar (www.domtar.com) is the third-largest producer of uncoated free sheet in North America and the fourth largest in the world. It is also a leading manufacturer of printing, publishing, specialty and technical papers.

MD: Why does paper crack?

DD: Unfortunately, there is no simple answer. Paper cracks because we bend the wood fibers beyond their bending strength. A fiber has certain flexibility, just like steel or plastic. When this flexibility is exceeded, the fiber breaks, producing sharp edges that protrude and are seen as cracking.

To help visualize it, think of a toothpick supported at each end. Now apply pressure on the center of the toothpick. It will bend. If the pressure exceeds its breaking strength, it will snap. Splinters or slivers will protrude out from the main body of the toothpick, although it's still not completely broken in half. This is what fibers look like when they break.

Wood fibers are like any other material. They can take compression and tension to a certain level, but beyond this level, they fail. Certain wood fibers are stronger than others, but they all have limits.

MD: Is it something specific about the paper's composition that causes cracking?

DD: Fibers are actually broken due to the force applied on them. When paper is bent against its grain, fibers are stressed at the point of the fold. (During paper manufacturing, paper fibers are aligned in a certain direction. If the paper's folded "against the grain," the fibers will snap.)

MD: Do certain paper types tend to crack more??

DD: The thicker the paper (higher caliper in paper makers' terms), the more likely it will crack. Coated papers tend to crack more than uncoated. However, this is because the paper's coating flakes and enhances the appearance of cracking. Paper cracks during the folding process only.

MD: Is there any rule of thumb a buyer or designer should know about to prevent cracking?

DD: Remember that paper cracks when it is folded against the grain. It does not crack along the grain. Designing a piece so that the folds match the grain direction will prevent cracking. This of course is not always possible, and that's when scoring becomes essential.

MD: Speak about scoring a little bit. I know that helps prevent cracking.

DD: Scoring is a process whereby paper is compressed or indented along a straight line. It must be done in the direction opposite the fold. The score should be wider than the thickness of the paper. Scoring reduces cracking because it reduces the bending force on individual fibers. On very heavy products like cardboard, double scoring is used to further reduce the stress.

MD: What else can a printer do to prevent cracking?

DD: Scoring is the best medicine, but other environmental factors are important. If the moisture of the paper is low, it will crack more easily. Sometimes it will need to be replaced. Low moisture in the printing plant will produce the same problem. Devices that increase plant humidity can correct this. Some plants actually spray water (mist) to increase the moisture at the point of folding.

Most people notice that paper doesn't crack much in humid summer temperature. So increasing the humidity is the obvious thing to do. If the relative humidity is below 35%, cracking is more likely to occur. To protect the paper moisture that is put in the paper by the mill, printers should avoid opening the packages until they're ready to use them. As soon as the moisture-proof barrier is broken, paper starts to lose or gain humidity, so keeping the paper in its original package or covered with plastic film is recommended.

MD: How much of an issue is a paper's weight, or its finish?

DD: The finish is not really important, but the caliper (thickness) is. Usually heavier papers are also thicker, and therefore tend to crack more.

MD: Solids present a potential cracking problem, don't they?

DD: It's a good idea to avoid putting ink on the fold, as any ink will enhance the visual defect. The darker the print, the worse the appearance. White paper doesn't show cracking as much. For instance, you can reduce the appearance of cracking on a paperback book by avoiding solids wrapping around the book end.

MD: Any final thoughts about cracking?

DD: Paper makers use different fibers for various grades of paper. For example, envelope grades are made with a higher percentage of long fiber (softwood), making them more flexible and increasing their resistance to cracking. So using offset paper to make envelopes increases the risk of getting envelopes that crack. My final word of advice? Use the proper paper for the job.

For more information, feel free to contact Denis at 613-938-4661 or at denis.doutre@domtar.com.

Chapter 7

SOME OF THE PEOPLE WHO MAKE IT HAPPEN

B uying printing? Make it personal. You need to like the people you're dealing with at the printing company or else it won't work at all. There are lots of different specialists in the industry, most of whom work behind the scenes. The sales reps get all the glory—but the folks in the pressroom and service and prepress departments deserve much of the credit, too.

This chapter takes a look at just some of the key people inside the printing industry.

Love Thy CSR

Customer service in a printing company is often a key determinant in whether a customer stays . . . or strays.

Customer service reps (CSRs for short) are the people inside a printing firm who take care of a customer once the salesperson has made the sale. How much (or how little) the salesperson manages each print job through the printing process varies from firm to firm and person to person.

Sometimes, CSRs are called Account Managers. Small print shops won't necessarily have designated CSRs, but rest assured, this work (servicing the account, hand-holding the client's job) is still being done, either by the sales rep or the owner.

Build a good relationship with your CSR from the beginning. Tell him or her how you want to be contacted, whether by phone or email or fax. Talk about how much information you need for each job in order to make your life easier.

Then make your CSR's life easier. Ask if your information is clear and comprehensive enough. Get forgiveness in advance for all the times you'll be barking orders at him or her over the phone. Take your CSR to lunch now and then.

Because it's tough being on the inside, where it seems like the salespeople are getting all the credit. CSRs juggle many customers and many jobs all at once, and as deadlines draw near and converge, they have frantic customers going crazy on the phones.

So be kind to your CSR. It's a tough job, servicing print customers. In my print-buying days, it was often my CSRs who found my typos, who came through with partial shipments, who had solutions for my many printing emergencies.

I probably didn't thank them enough. Make use of my hindsight and thank your CSR today.

Go for Broker

Pity the print buyer who dismisses all print brokers as unnecessary, unqualified . . . or worse.

I didn't use brokers when I was a corporate print buyer. Back then, I was cocky enough to think I didn't need them. I couldn't see any value they added to the process. I worked with printers through their salespeople. Period. End of story.

But since then, I've come to realize how valuable a good print broker can be to a buyer, especially a buyer who doesn't have a lot of experience, as well as one who purchases so much printing, of such a wide range, that he or she can't possibly keep up with which local printer does what. Add to that the reality that printing is getting more and more complicated, thanks to digital technologies, and you begin to see why print-buying requires ever-expanding technical knowledge.

Remember, commercial printers usually specialize in certain types of products, based on their equipment. They have alliances with other printers to handle what they cannot. This is called outsourcing. It's done all the time, whether or not the buyer knows it.

So when it's put in this perspective—why is working with a good broker any different?

Mind you, anyone can call himself a print broker, so buyer beware.

In my mind, the best brokers are experienced in the manufacturing process: they've worked in print shops for a long time, not just on summer vacations. They have a history in the region. They have a client list and references that demonstrate their credibility. They're responsible for everything on a job after you give it to them: selecting paper, scheduling, showing you proofs, troubleshooting, doing press checks, monitoring delivery, the works.

The best brokers can also offer you suggestions for saving money. They know all about paper and ink and the dozens of presses out there today, both offset and digital.

In short, they're experienced, technical experts who represent many printers. They are agents.

So don't be so quick to dismiss all brokers. Check them out as you would a print salesperson. Interview them, find out how much they know about the manufacturing process. Get a sense of how deep their knowledge is, and determine where their responsibilities begin and end.

If the term print broker has negative connotations, maybe it's time to change the name. "Print agent" has a nice ring to it.

Graphic Designer vs. Desktop Publisher

Many people think that "graphic designer" is synonymous with "desktop publisher." It's not so. Every graphic designer is (or should be) a desktop publisher, but the converse isn't true. "Desktop publisher" generally refers to someone who has experience in creating digital documents with layout software like QuarkXPress, PageMaker or InDesign.

All too often it just means someone who's handy with a Mac.

A "graphic designer" is someone who has earned a professional degree in design. He or she is educated in the art and science of visual communication, encompassing typography, signage, identity systems, packaging, and computer interface design. Designers know about color perception, proportion, type design, and various design methods and applications. Desktop publishing is but a kernel of what they can offer clients.

So be thorough when you interview candidates to design your materials. There's a lot more to creating a brochure or corporate stationery than choosing a pleasing typeface and nice paper.

Specifications Reps: The Paper Pros

What would the printing industry be without paper? The Internet?

Like marriage without love, Desi without Lucy, or crème without the brûlée—printing without paper is inconceivable. And no fun at all. Luckily, printing and paper still cut a mighty fine figure together.

More than anything else, paper is tactile. It's all texture, color, and finish. And as much as there is to know about printing, there's just as much to know about paper.

How can you learn about current paper choices, or what paper might be best for your project, or what new paper is being produced by what mills, and a host of other critical paper issues? By working with a specification representative from one of several paper merchants or paper mills.

Remember, paper mills make paper. Mills sell paper generally through paper merchants, who in turn sell paper to end users and to printers, too.

Specifications reps are paper specialists who work for either merchants or mills. Their job is to meet with buyers and graphic designers and recommend papers, discuss trends, and supply dummies and swatch books (paper samples).

For the most part, no one merchant or mill carries every brand of paper. But spec reps, as these specialists are called, know all about available paper, even if they don't carry a particular brand.

What's the value of specifying paper through a spec rep? Well, they give you personal, professional advice and service, which helps you choose the best paper for a particular job. They can get dummies made up for you. Good spec reps give programs on paper for their customers—they'll come to your corporate office or design studio and tell you all you want to know about paper.

Spec reps are a wealth of resources for the end user. They provide promotional as well as educational material for their customers.

They're like a traveling troupe of paper pros, these folks, bringing you all kinds of beautiful printed materials and the latest swatch books to help inspire your choice of paper.

If you take their paper advice, just remember to tell your printer to use them when specifying paper for that particular job: it's how they're compensated. (But I 'spec you knew that already.)

Portrait of a Modern Printer:
"Knee-Deep in W-2's!"

Picture the owners or presidents of printing companies ... how did they end up there?

Did they learn the trade as young apprentices? Did they study the craft in grade school or high school? Were their fathers and grandfathers printers? Until about a decade ago, chances are that these answers were all correct.

Today, that traditional, almost romantic, image of the making of a printer is fading.

I began to notice the change a few years ago. While interviewing company presidents for articles in *New England Printer and Publisher*, I saw a new breed of CEOs. Several owners didn't have any printing in their blood. What was happening here?

I searched for some insights and found them in a conversation with Mr. Gardner LePoer. Gardner is the Executive Director of the Museum of Printing in North Andover, Massachusetts. He dates his entrance into the printing industry to 35 years ago, when he was a newspaper editor. In addition, he served as former chairman of The Montachusett Regional Vocational Technological School District, which serves students in 18 towns in central Massachusetts.

Years ago, the printing trade was taught on the job, in print shops everywhere. It was also taught in public middle schools and in vocational-technical schools. A new printing apprentice would work his way up through the ranks, learning to operate bigger and more complex printing presses as he progressed. Print shop owners learned their trade from the ground up.

Beginning in the 1960s, according to LePoer, the traditional education of a printer began to change. Educational funds were advancing regionalization, causing public school systems to shift the instruction of printing from high schools and middle schools to large, regional vocational schools. This narrowed the opportunities for many young people, as it decreased the exposure to and appreciation for the printing trade. As a result, says LePoer, many individuals who work in the industry today have little or no exposure to printing until they walk in the door.

Printing education in vo-tech schools has changed as well. Why? There are two reasons, says LePoer. The first is that today, vo-tech schools are 50% female. The printing business doesn't sufficiently attract women, particularly in manufacturing roles. Second, the focus of printing programs in vo-tech schools is on prepress. It may not be universally true, but it has certainly contributed to a lack of skilled pressmen.

Coursework in digital prepress technology sells. Let's face it: it's the "sexy" side of printing. And you don't have to get your hands dirty.

So who is running printing firms these days? More and more, says LePoer, it's businessmen and women who are "knee deep in W-2 forms," as opposed to paper from a printing press. Certainly, men and women who run the printing and finishing equipment still populate print shops all over the country. But today's printing companies require IT expertise and all sorts of computer-savvy and management professionals.

Literally and figuratively, press technology is becoming a cleaner and cleaner business.

One thing's for sure. This business is "not your father's print shop" anymore. Printing is high-tech and cutting-edge. It's as much about Web publishing (the "www" kind, not the press kind!) and digital asset management as it is about ink on paper.

The "modern" print shop has added high-tech muscle and global reach to its products and services. That's not a bad thing, considering how fierce the competition is. But what about the original craft? How will the younger generation learn about it?

Gardner LePoer speaks for a lot of us when he says that he'd like us to recapture an appreciation for the extraordinary craft of printing. It's not as easy as it seems.

For information about the early days of printing, check out the Museum of Printing, which is open to the public. It houses a massive collection of printing artifacts and machinery, including printing presses, typesetting equipment, bindery equipment, and early computers.

For information about tours and other offerings, contact LePoer at info@museumofprinting.org or visit www.museumofprinting.org.

Chapter 8

Going Deeper into the Printing Process

T he longer you work with printers, the more evident it becomes that this business is technology-driven. Printing something well requires a lot of work and decision-making before the job prints. Plus, the advances in digital technology have improved and sped up the prepress stage of every print job.

Why should you care? Because printers work hard on your jobs behind the scenes. Often, that's what takes the most time!

This chapter takes a look at some of those "behind the scenes" printing technologies.

Get Ready for Makeready

Some people still confuse laser printing with real printing. My new laser printer was up and running in minutes. Consumers might expect their commercially printed jobs to be done almost as quickly.

"Don't they (commercial printers) just push a button?" I'm asked. "Well, no, they don't," I reply.

So today I want to talk about makeready.

"Makeready" is a common printing term that describes all the steps that are required to set up or prepare a printing press before printing a book, magazine, or indeed, any printed product.

Makeready steps include the washing up of the press from the previous job, changing of ink, and plate removal and re-plating for the next job. You don't see a press sheet until after the makeready stage.

If you're printing your job in special colors, this is an extra makeready step, as the press will have to be cleaned up before and after your print run.

In the case of a web press, makeready may also include webbing up of a new stock, which means putting a new web or roll of paper onto the press.

Sheet-fed makereadies are quicker than those for a web press, just as makeready for a single-color press is quicker than that of a four-or five-color press.

Makeready can also include matching printed results to a customer-approved proof.

Extra copies of a job already on press are significantly cheaper due to the absence of any further makeready charges. Bear this in mind when placing your initial order.

Theoretically, digital printing doesn't require a makeready, says Mark Miller of the Japs-Olson Company (www.japsolson.com). The reason, he adds, is that you receive a digitally printed component from the first file sent to the press—much like a laser printer in your office. However, some makeready time and materials are required in digital printing to produce the best printed materials.

Jamie Bradley, Print Management Director with Fusion (www.fusionpromo.com), had additional comments worth noting.

"With digital toner printers," wrote Jamie, "printers will usually run a test sheet before outputting the entire job on this type of press (this might be the proof you see). On high-quality machines (like the Indigo or the NexPress), there's typically a higher start-up cost to cover preflighting files, proofing, and setting up the equipment.

"With digital offset presses (like the Heidelberg DI), they are still putting ink on paper, and printers need to run 25-100 sheets or so to balance color and check register. Start-up costs are a little higher, but the cost for extra copies is lower than toner-based machines."

When you talk about makeready costs on offset presses, there are several factors to consider, added Jamie, who went on the explain:

— Time is money.
— Different presses have different hourly rates, and setting up more plates takes longer.
— Newer presses are up to 75% faster to set up than most old presses.
— Paper waste and cost of the paper you choose (old presses waste lots more paper).
— Paper types vs. complexity of job (lightweight text stocks stretch on press and take longer to register).
— How fussy you are about such things as color, register, consistency, and double-checking for typos or hickeys or missing graphics. (That's why printers want to know when they quote if you will be there for the press check.)

Getting a press ready for your job takes time and preparation. Printers are not simply accepting your file with one hand and pressing an "OK to print" button with the other. Keep it in mind the next time you're tempted to hold up a hoop for your printer to jump through.

Scanning Basics

"Why not do a simple column on photo scanning tips for new designers and desktop publishers?" I thought. Simple? Hah! The more I investigated, the dizzier I got.

If you need to scan in photographs for a printed piece and you're new at it, do some or all of the following:

1: Talk to your printer or prepress house first.
2: Read Frank Romano's *Pocket Guide to Digital Prepress*.
3: Read *Getting It Printed* by Mark Beach and Eric Kenly.

Why do you even need to scan photos? Because they're continuous tone images and need to be converted to dots for printing presses. That's what scanners do: convert continuous tone images to digital ones. Black and white photos get converted into halftones, which are dot patterns. When color photos are prepared for printing, they're called separations. (Halftones and separations are two very common industry terms. Remember them.) Each of the four process colors (4-C, or CMYK) comprises lines of dots. Each color is lined up at a different angle to prevent the printed piece from looking like a TV screen.

Back to scanning 101. There are many kinds of scanners out there. I spoke with an expert who knows more about scanning than I can ever hope to. John H. Fetter is a magician when it comes to color. He started out in the printing industry in 1950. He worked for Lebanon Valley Offset in Pennsylvania for 45 years. Today, John's an independent consultant to printers and corporate end users alike. His specialty is doing proofing qualifications for printers and designers as well as press OKs. Contact John at (717) 867-2791 or at john.fetter@verizon.net.

Run-of-the-mill slide scanners scan 35mm slides. But the

two main types of scanners are flatbed scanners and drum scanners. Flatbed scanners are relatively low-end and are considered all-purpose scanners. The high-end scanners are drum scanners. The better the scanner, the more sensitive it is to reading (or matching) the colors in the original image. The drum scanners are still superior, said Fetter. They can see into the shadows better and give you much better quality for high-end work. They also tend to scan at a higher optical resolution than flatbed scanners.

In scanning an image, you're controlling four things, writes Romano: resolution (usually measured in dpi or dots per inch), tonal range, color balance, and sharpness.

Color scanning is more critical than black and white. Some colors are more important than others, like skin tones. To get the absolute best color work possible, prepress professionals often do production work under color-correct lighting conditions.

Color images created on a computer must be converted from RGB (Red, Green, Blue) to CMYK (the 4 process colors used in printing, Cyan, Magenta, Yellow, and Black). Scanners can convert "on-the-fly," said Fetter, as opposed to you using software to convert RGB to CMYK, but he also noted that many scanner operators don't have the preferences set up properly on their scanners. He often prefers to begin work with (or color correct) the RGB image rather than the CMYK image, since it's the purest image.

I sought scanning tips from local designers. Here's what Linda C. Lyons, Senior Designer for Champagne/Lafayette Communications in Natick, Massachusetts (www.chamlaf.com), said:

"To get the best result, your image must be scanned at the highest quality possible. Although an office flatbed scanner might be OK for some things, for high-quality printing, provide the original transparency (or print) to the printer to scan on professional equipment. Color corrections can often be done at this time as well."

Ken Hablow of Ken Hablow Graphics in Weston, Massachusetts (www.khgraphics.com), added these tips:

"Scanning for print is better at a 300 dpi resolution. This will depend on the line screen of the negatives used to make the plates.

If you start with a poor original, your scan will output poorly. Photo editing software can only do so much.

You shouldn't scan something that has already been printed. Anything printed is made up of dots. You are then scanning dots to produce yet another image of dots. This causes a moiré effect. Most scanning software will "descreen" a printed image, but it is not ideal. (Fetter notes that this will work, however, if the output size is smaller or if that's all you have to work with.)

If the final output will be larger than the original, then the resolution of the scan needs to be increased so the finished size is still 300 dpi. This is best done in the scanner, not with photo editing software.

If the original is a transparency, the best result will be done with a commercial drum scanner, not a desktop scanner. It's costly but necessary for high-quality output, especially if the image is going to be used larger than 4" x 5."

This is just a hint of what scanning photos is all about. Frank Romano's final suggestion for successful scanning is to check with your printer or prepress service. That would be my FIRST suggestion. As I questioned John Fetter about whether you could do this or that when scanning, he'd often say, "Yes, you can—but you don't want to!"

When it comes to scanning, there's nothing black and white about it.

The Scoop on Prepress

Prepress is one of those very common terms used in printing that's not often explained to newer buyers. Since it's an integral part of every printed job, I think it's time to get the basics "out on the table." So I enlisted the help of an expert.

David Cohen is Operations Manager for The Smith Print in Norwell, Massachusetts (www.smithprint.com). I interviewed David about the prepress process so that print customers, including graphic designers, can gain a better appreciation of this process. Here's what he had to say:

MD: What are the main parts of prepress, exactly?

DC: The overall job of prepress is to take the customer's application files (usually QuarkXpress, PageMaker, Photoshop, Illustrator, Freehand, etc.) and prepare them to be imaged at high resolution to produce printing plates. Since the application files are usually set up as single pages or reader spreads, the prepress department also has to layout the job according to how it will run on the press sheet. Proofs of the job are prepared and approved by the customer, showing the layout and color match that will be used at press.

MD: Is preflighting a subset of prepress?

DC: Preflighting is the process of evaluating the customer's application files to ensure that they will output correctly. This usually involves checking for correct fonts, high-resolution images, correct bleeds and trims, and application settings that affect final output.

MD: Do the majority of offset printers perform the same steps in prepress? I'm wondering if there are "prepress standards" in the industry.

DC: The necessary and "standard" steps include ripping, trapping, and imposition. All prepress departments at commercial printing companies perform these operations. At Smith Print, we also perform many quality checks to ensure optimal reproduction. The "higher end" prepress departments are those that have set up their operations to anticipate press reproduction. This includes things such as linearization, color management, and press "fingerprinting."

MD: Can you define RIPping, trapping, and imposition?

DC: The RIP (raster image processor) is a device that can interpret the page layout information for the imagesetter or plate setter. Trapping generally refers to the way colors overlap each other so that there is no gap between them on press. Imposition is the process of taking the customer's document and laying it out according to how the job will be manufactured.

MD: Where and when does prepress begin and end? Does it take a specific amount of time to complete?

DC: Prepress begins once the designer completes the customer's files. The first step is to review the designer's work and understand what he or she has built in the files. The time it takes is directly proportionate to how the job was created by the designer.

MD: Does it include makeready?

DC: No, makeready is the process of getting the press ready to print a job.

MD: Am I right in assuming there's a lot less prepress in digital printing?

DC: No. It's just different. Software provides designers with many options for building jobs and most designers have their own ways of doing things. This lack of standardization in the industry often makes the prepress operator's job more difficult. The digital workflow has significantly streamlined the process. However, it's important to realize that most presses still require plates, and the issues of color match and registration still exist on press. Even though we now do it digitally, prepress still has to address these concerns.

MD: Does the amount of prepress on any given job depend on how complex a job is?

DC: The amount of prepress is directly related to the way the job will be printed and manufactured, not what the final product looks like to the average reader. For example, you might have a solid color pocket folder that looks simple. But in order to get it to print that way, to lay down smooth, clean solids on an offset press, you might run additional layers of color to boost the solid and then add a varnish or coating. So, the complexity of a job is based on the requirements of the entire manufacturing process.

MD: Are there things that customers and/or designers can or should do with their job to make the prepress stage go more smoothly?

DC: Generally, I have found that the longer a designer works on a job to build it, the more difficult it's going to be to output it correctly. This usually happens when a designer is not sure how to set something up and begins to experiment. Most, if not all designs, can be accomplished in very few steps if you know how to build the job. Sometimes experimenting is necessary, and it sometimes means starting over once you've figured it out. If you submit your "experimental" file to prepress, who knows what will happen?

MD: Any final comments?

DC: One good resource for designers is a publication called "GRACOL," General Requirements for Applications in Offset Lithography. It's the best resource that tries to establish production standards in the digital world. It's available through the IDEAlliance (International Digital Enterprise Alliance, formerly the Graphic Communications Association). Call 703-519-8160 or visit http://www.idealliance.org. It's also available at http://www.gracol.com.

Digital Printing in Plain English

Digital printing is everywhere. You probably know the advantages of going digital—great for short runs (up to 5,000 copies, generally speaking), very fast turnaround, lets you send digital files across the miles to "distribute and print" as needed, as opposed to the traditional "print and distribute" model, eliminates waste (you only print what you need). These are the key ones.

This technology is relatively new (a few decades old) and very hard to pin down, since the equipment changes rapidly, and digital capabilities just keep getting better and better and better.

Do you know what's different about a job printed digitally? I went searching on the Internet for a plain-English explanation, but came up empty. Many sites presume that visitors already know what digital printing is. They talk about benefits but don't bother to explain the basic technology.

Okay, some people could care less about HOW it's done—they just want it done NOW and want it done as cheaply as possible. But for those of you who are a little more curious, this one's for you.

WARNING: This isn't a look at hardware or software. Digital presses all have different capabilities and limitations, which keep changing.

Rather, this is a simple explanation of digital printing to give businesspeople a better understanding of how it works. If you're working with a printer who does digital, and you want more detail about how a particular press works, just ask. (I know of very few commercial printers who do NOT offer digital printing these days. If they've not invested in their own equipment, they at least have an alliance with printers who do digital.)

Offset lithography is a mechanical process that uses ink on metal plates to transfer images from a rubber blanket (a roller) to the paper, applying tons of pressure to do so. Printing plates are the image carriers in this traditional method. The plates are imaged once and used over and over again. Offset lithography is the most efficient printing process for long runs. Offset inks are oil-based, and they dry in the air, after the printing's done.

In digital printing, no plates are used. It's not a mechanical process; it's electronic. A new image, made from your digital file, is applied to the paper every time. So the 1000th sheet should look exactly like the 1st sheet. Each printed sheet is an original, not a copy.

Digital printing uses laser technology or LED (Light Emitting Diode) technology to expose a drum to the image, and then transfer this image digitally to the paper.

Most digital presses are toner based. Some use dry toner; some, liquid. Toner has to be heated up in order to dry. Some presses have digital "front ends" only (for the platemaking part), but use conventional presses for the printing function itself.

Some, but not all, digital presses can do variable data printing. Offset presses cannot do variable data printing.

With digital printing, you can print exactly the amount you need. You waste less paper. You can change your content/ images "on the fly," so your message is always current. You can get great full-color printing on digital presses. The quality gets better all the time.

There are limitations in digital printing, however. Paper choices, finished sizes, and certain processes (such as spot varnishing) are just a few.

When designing a digital job, I beg you, talk with your printer first. I've been told umpteen times that you MUST design for the specific equipment your job will run on.

That's a quick look at digital printing from 30,000 feet. Trust me, it's just the tip of the iceberg.

Top 7 Reasons to Use Digital Printing

Digital printing is the newest printing process—and one that's stood the industry on its head.

Unlike traditional (offset) printing presses, digital presses are made to produce short-run (1 to 5000 copies), quick-turnaround (under two days), 4-color jobs. These new presses are made by such companies as Agfa, Canon, Indigo, Heidelberg, and Xerox.

In traditional printing, your job goes from your desktop to a digital proof to an imagesetter (for making film) to another proof to a platemaker to the printing press. Phew!

But if you "go digital," your job goes from your desktop to a proof to plates to the press—or even directly from desktop to press. Zip-a-dee-doo-da!

You can benefit from using this technology in many ways. Here are 7 good reasons to "go digital" in printing:

1. You can print as little as a quantity of one. Seriously, your print runs can be low, even for full-color printing, and the costs won't make you blanch.
2. You can print something in time for that important meeting or trade show next week. Lead times are much shorter when you go digital.
3. You can print different versions for different markets, testing their effectiveness before you settle on one look.
4. You can personalize your materials by taking advantage of the latest subset of digital printing, called variable data printing (VDP). Every piece has different content, because VDP marries a database (yours) with a design template.
5. Your salespeople or executives can have their own version of sell sheets or other printed materials, tailored specifically for their own clients or prospects.

6. You can change the content more often with digital printing. So print 2000 brochures every quarter, instead of 8000 brochures every year—and keep your information current for a change.

7. You can kiss those unused printed pieces goodbye! With digital printing, you can reduce or even eliminate waste.

So if you need just a few or a few hundred copies of your 4-color newsletter, marketing brochure, trade show packet, or business cards, go digital.

Is digital printing perfect? No. Limitations include paper size and color choices. And not every printer offers it. But the technology improves every year. Give it a try!

VDP: Where the Printing Is Mine and Mine Alone

The newest kind of printing has an ugly name: VDP. It stands for Variable Data Printing, and it is a kind of digital printing. Although VDP's been around for several years, it still exists below the radar for many folks.

It's also called Customized Printing, Personalized Printing, One-to-One Marketing, and Variable Information Printing. A mouthful, one and all.

In plain English, VDP is the ability of certain digital presses to merge data from a database with a design template, and then generate unique content on the documents that are printed. The emphasis is on "unique content."

There are several kinds of VDP, from simple to sophisticated.

If you've ever done a mail merge, where you insert address information into a letter template so you can mail the letter to a list, that's simple VDP.

If you sell products nationwide and develop different versions of your sales materials for different markets, that's VDP.

If everyone in your household receives mailings from the same source—and all that's different on each piece is everyone's name, printed somewhere in the copy, that's VDP. Next up is customized printing, a higher level of VDP that requires more programming skills. If you visit a car manufacturer on the Internet and request a product brochure based on your preferences, what shows up in the mail will be a brochure that you "built" online. The printed piece is created to reflect your personal preferences. That's definitely VDP.

Financial statements are types of VDP. They're more sophisticated these days and likely to include colorful charts and graphs that reflect your personal financial data.

The coolest VDP application is the toughest, technically speaking. Here's an example: You do a lot of shopping at a certain retail store. Sometimes you shop at their online site. They're keeping track of your shopping preferences by collecting data (a bit creepy to me). One day, a catalog arrives that only includes the stuff you're interested in. Or a packet of coupons arrives from a store you frequent, offering discounts on items you've bought in the past.

VDP is very much a specialty, and few printers can do it. Although unit costs are higher than for offset print jobs, VDP increases the effectiveness of your marketing message. It's targeted marketing. According to the trade pubs I've read, the response rate with VDP is a whopping 24% to 34%, as compared with low single-digit response rates with mass marketing materials.

What does it take? A marketing plan, good data, strong programming skills, a designer who's done VDP work before, and a printer's who's experienced in it as well. Once you put this dream team together and make sure all of the pieces fit, you test, test, and test again. Start small. Don't be impatient.

Earlier this year, an issue of *American Printer* magazine was sent to all 82,000 subscribers in distinctive wraps. Each wrap was personalized on the front and back with our first name and last name, and different photos and illustrations as well (from a batch of 52 images). So my cover is mine and mine alone. Tres cool.

CTP—What Exactly Is It, Anyway?

Printing technology is changing, especially at the front end of the process. One newer technology is CTP. (Printing is filled with confusing acronyms, but that's another story.) To help explain CTP, I interviewed Duncan Todd, CEO of Champagne/Lafayette Communications, a printing and marketing firm in Natick, Massachusetts. Champagne/Lafayette has been 100% CTP for several years already.

MD: What does CTP stand for?

DT: Computer to Plate. It's a platemaking technology that allows us to bypass film and image printing plates directly from digital data.

MD: Is CTP the same thing as digital printing?

DT: No, it isn't. CTP utilizes digital data to image printing plates, eliminating film from the process. Digital printing uses digital data to image directly on the printing device. Direct Imaging incorporates CTP platemaking on the press itself. Once the plate is made, Direct Imaging presses run using the same technology as standard offset presses. Examples include the Heidelberg Quickmaster DI and Speedmaster DI presses, Man Roland's DICOPress, and the Komori Project D press.

MD: In a nutshell, how does CTP work?

DT: CTP plates are made separately from the press and must be mounted on the press.

MD: Is CTP technology independent of a specific press?

DT: Yes, the CTP platesetter is independent of the press. CTP plates can be used for any press compatible with the plate sizes that the CTP device can make. We use one device to make plates for a duplicator size (12 x 18"), 28" and 40" presses.

MD: If a printer has CTP technology, does he or she run every job that way?

DT: In our case, we run every new job and most reprints using CTP technology. Champagne/Lafayette was the first commercial printer in our area to eliminate film totally from our production process (with the exception of a few very old jobs standing in film without changes for which we do not have electronic files). Some commercial printing companies have continued to run parallel systems. Such companies may prefer to use film to update jobs that are standing in film rather than convert the entire job to a "fully-imposed" workflow for CTP. Some advertisers continue to supply film to publications so that they can "lock" the content of the ad, although this practice is changing, especially with the advent of PDF workflows that offer the same insurance against the printer inadvertently introducing error into or changing the content of the ad.

MD: Does CTP cost more or less than traditional offset printing?

DT: A well-honed CTP workflow can reduce a printer's costs. The cost of the film is eliminated, along with the cost of properly disposing of the film processing effluent and scrap film. Press makeready times also decrease because of better register on the plates and the sharper dot structure. The printer has increased his capital costs for this new technology while lowering his labor costs. The savings are a function of this tradeoff. However, if a printer does not have control over his workflow and rework, he will just be throwing away more expensive bad plates in place of discarding bad film.

MD: With CTP, do clients see proofs quicker?

DT: This is a scheduling question. As a practical matter, getting to digital proofs can be a little bit faster compared to the time required to make proofs from fully imposed film. It is definitely faster than the older system of outputting individual pages that are then stripped manually. Depending on the format size and speed of the digital proofing device, some analog proofs can be made in less time than digital proofs. Even though the process is digital, it's not instant.

MD: Does CTP mean a client sees a certain type of proof?

DT: CTP incorporates digital proofs rather than analog proofs. There are several types of digital proofs: laser printer proofs (appropriate for content, position, color breaks, cropping); digital ink-jet proofs (high, medium and low resolution), appropriate for all the previous plus back-up and folding; dye sublimation proofs (all the former except back-up and folding, plus color approval); and digital halftone proofs (all the same as dye-sub, with the added benefit of using halftone screening, usually the same screening as used on the plate).

MD: Does CTP mean that a firm has a totally digital prepress department?

DT: To employ CTP, a printer must have a complete digital workflow, including imposition and digital proofing. It does not necessarily mean that CTP is the only workflow. A company may use a fully imposed film workflow (all digital until the proof stage) or traditional stripping of individual pages or groups of pages.

MD: What other benefits does a client see with CTP?

DT: The electronic files used to make the proofs are the same ones used for imaging the plates. If the proof is approved with no corrections, no additional processing is required other than sending the ripped file to the platemaker. If there are corrections, less time is required between the proof return and platemaking because there is no film to image and strip.

MD: What are the downsides?

DT: There are very few downsides to the client. The biggest difference is the substitution of ink-jet proofs for the traditional salt print/blues/Dylux proofs. Because the proofs are not analog proofs, the ability to determine print quality or check screen values on the proof is lost unless one uses the more expensive contract proofs. But one gain is the ability to check color breaks.

MD: Need the client know anything about CTP when preparing files?

DT: If the client has been creating files for fully imposed film, there is no change going to CTP. The basic rule still applies: check the file before it is sent to the printer to make sure the file's set up correctly and that all elements are included and created in the proper format. Remember, if it doesn't print on your laser printer, it's not going to make it through a printer's RIP without a lot of extra work by the printer (read: extra expense to the client), if at all.

MD: Does CTP affect the print quality of a job?

DT: CTP improves the quality of the printed product compared to film-based platemaking. It eliminates plate problems such as broken type, dust problems, and poor negative-to-plate contact (halation) that plague conventional platemaking. CTP significantly improves register on press. Dot gain is significantly reduced, resulting in sharper printing and stronger color.

MD: Any thoughts about the future of CTP?

DT: As the cost of CTP systems comes down, more printers will adopt the technology. Making the transition to CTP is much more a workflow challenge than a technology challenge. The time is approaching when lithographic film will retire to its place of honor next to rubber cement, rubylith overlays, and foundry type.

There you have it, a clear explanation of CTP technology and how it benefits customers.

Finally! Basic Q & A's
for Consumers about Print on Demand!

The printing industry hasn't done a great job of explaining POD to consumers, although they insist that this fairly new technology is going to change buyers' lives. Too many basic questions about POD go unanswered. Until now.

I've enlisted the aid of an expert to provide information that will help everyone curious about POD, or Print On Demand.

Barry Reischling is the founder and president of RPI (Reischling Press Inc.) in Seattle, Washington. RPI's web site is www.rpiprint.com. I asked Barry a series of questions about POD that I think most consumers would ask. Here's what he said:

MD: I've been reading about POD since the early '90s, it seems. What exactly is it, and how does it differ from other types of printing?

BR: POD means Print On Demand. Simply stated, digital POD printers offer buyers the economic benefits of speed by printing shorter-run high-quality B&W and full-color jobs faster than can most other printers. When one wants to avoid the space hassles of print inventory, or needs immediate inventory replenishing, or has a need for more print distribution, they should consider POD.

At RPI, we endeavor to print immediately for same-day or next, or 2-day or 3-day. In most cases, POD jobs reprint from stored original print files.

MD: Does POD refer to specific printing equipment, computer technology, or a new process for getting things printed?

BR: Any printer can say they do POD. But most legitimate POD printers must first have expertise in and understand the significance of POD market applications. They'll need a fast T-1 internet connection to speed up transfer times of huge color files, extensive computer front-end systems, complete software libraries, automated prepress production, and for back up, a minimum of at least two high-speed B&W and digital color presses, not copiers. For production control purposes, their own in-house bindery including fold, saddle-stitch, and perfect, is also mandatory. But most critically important, they must feature a standard 24/7 work ethic.

MD: Are the terms POD and digital printing interchangeable?

BR: No. POD and digital printing do not mean the same thing. Many printers feature digital press technology. However, they are not POD printers. POD printers have the required press and bindery technology to turn jobs much faster than the majority of printers. Their POD service and 24/7 operation should be a normal part of their every day disciplines.

MD: What kinds of jobs are best suited for POD?

BR: Our typical POD jobs are soft and hardcover books where publishers don't want the storage hassle and expense of inventory. Other products include appellate briefs, legal/financial documents, SEC corporate disclosures, venture capital private placement/ disclosures, full-color marketing pieces, full-color real estate listings—or almost any job under severe time constraints, printed to meet just-in-time deadlines. Many of our POD jobs are short-run variable image where we can program different images to be printed for each successive impression.

MD: Is there an optimum print run for a POD job to be cost-effective?

BR: Not necessarily. We even do one-offs periodically. Unit cost effectiveness is relative, so optimum print runs are seldom an issue. Usual POD runs are from 50 to 2,000 copies.

MD: Are there other jobs/circumstances for which print run is not critical, yet POD makes business or marketing sense?

BR: Yes. We do much in the way of pre-published books. In the trade, they're referred to as galleys. But in this case, a galley is a sample of an actual book, completed to specifications, bound and trimmed to size, many with full-color covers. Publishers want to see POD short-run books in final form right now. Their editors are paid to read and judge whether that book is worth more POD runs to further test market acceptance, or a full web publishing run. POD runs are usually for new authors. We also POD full-color catalogs for design, color, layout, and product position issue critiques, prior to full web production runs.

MD: Does a customer need special software for a POD job?

BR: 99% of our POD work comes to us via email in Adobe's PDF format, with separate source files for color. PDF is today's printing standard. We always email back a page-imposed composite PDF proof for an OK, prior to printing.

MD: Is turnaround faster?

BR: POD is always faster when jobs are in PDF. Benefit? We never have to guess what's required. We don't have to tweak Word files or mess with wrong fonts.

As a rule, digital POD press runs are completed more than twice as fast as they would be using conventional offset lithography. Where offset requires negatives, plates, press makeready, and ink drying waits before bindery, digital POD requires none of the above. But for quantities above 3,000, and where price is more important than turn time, offset may offer better value because offset's longer-run unit costs are lower.

MD: Can one make changes easily when reprinting a POD job?

BR: Yes. If changes are minor, we make them on print files recalled from storage. If major, we ask customers to do their own text editing.

MD: I think of a Xerox DocuTech as POD equipment for black and white jobs. Are there others?

BR: We feature four digital Docutech 600 dpi B&W variable image 6180s, and three digital Docucolor full color 400 dpi variable image 2160s. There are similar competitive machines on the market. We've looked at all of them. Spot color is done everyday on the 2160s.

MD: How good is the quality for color work? Is quality dependent on machine or operator or the customer's file?

BR: Output quality is dependent on all three—the machine, the operator, and the integrity of the file. Without a glass, it's impossible for even the most severe chromo-critic graphic designer to discern the difference in quality between our 2160 full-color product, and that same job printed by conventional offset. The practical difference? Our short-run 2160 full-color jobs are completed in a fraction of the time and at half the price.

MD: Are there paper or size limitations with POD?

BR: Yes. We cannot print press sheets over 12.6" x 19.2." Bleed maximums are about a quarter of an inch under this size. We have few text-weight paper limitations. Although our 2060 will print 10pt Kromecoat, there are many additional cover-weight limitations.

MD: Are there easy guidelines for knowing if your job should print POD vs. offset? How about page count—how much of a factor is it?

BR: Ask yourself, which is more important, time or money? Meeting the delivery date or job cost? Most of our POD jobs happen when time is more important. Better POD candidates are short-run, multiple-page, perfect-bound jobs. Many of our POD quantities are under 1000, but some have page counts exceeding 500. For single-page jobs, offset may be a better value.

MD: How does a consumer find a quality printer who can do POD . . . other than contact RPI, that is?

BR: I recommend the search engine Google for buyers looking for POD printers. Enter "Print On Demand" or "On Demand Printing" as the search subject. Many qualified POD printer-site ads will pop up to the right of your screen. Clicking on them will bring you directly to their Web sites.

How Not to Weave a Tangled Web

If you're under the age of 30, you might assume that "web printing" has to do with the Internet. Wrong! A web printing press is one massive piece of equipment that uses paper that comes on rolls (or webs), as opposed to sheets.

Web presses are built for high-speed, high-volume printing. Rob Francis of Transcontinental Bayweb in Ontario, Canada, made the following four points about web printing:

1. Generally, web presses have more dot gain than sheet-fed presses, so your scans should be prepared accordingly. (There are some waterless webs that can rival a good-quality sheetfed press for gain and sharpness, says Bob.) Your web printer can provide guidelines, as can the SWOP web site (Specifications for Web Offset Publications, visit www.swop.org).

2. Web presses can't print as much ink in the shadows of a photograph as a sheetfed press can. If too much

ink is used, you'll lose the details in these shadows. Talk to your printer ahead of time.

3. Every web press has a maximum printable image area. Any image that goes beyond this area will not print. Beware!

4. Web presses come in a variety of sizes and capabilities. Some can do inline gluing and trimming, special folding and perforating. Others just lay ink on paper. So designers and print buyers need to know which press a printer has before designing the job.

In short, check with your printer before sending in a web job. Good communication will save time, money, and aggravation.

Newspaper Printing...
A Process of a Different Color?

Planning on running a 4-color ad in a newspaper soon? A client who's a newspaper printing expert passed along these helpful guidelines to ensure your success:

Do some homework before having your repro or film made. First, be aware that dot gain for newsprint is at its extreme. Request your printer's technical specifications for your prepress vendor.

Newspaper presses are limited in the amount of ink they can carry. Once the ink hits the paper, it's dried mainly by absorption (unlike magazine web printing, which is dried by heat). Consider the TAC (Total Area Coverage, aka Total Ink Coverage). TAC is the total of all four ink densities in shadow or saturated areas. As a rule, the TAC for newsprint is around 230%—the lowest of any paper.

Newsprint registration (the matching up of two or more

graphic images so they're perfectly aligned) isn't as accurate as magazine presses. So avoid small, reversed-out text or logos, especially if the background color is made up of two or more colors.

Always take the paper into consideration. Newsprint presses have a much smaller "gamut" (the range of colors they can reproduce) than other presses. Some corporate colors won't be reproducible. Always get a proof made on the paper that's being used.

Speaking of proofs—you must sign off on a proof that reflects the final printed product. Forget about supplying the newspaper with a proof on glossy, white stock. Supply a proof that's an achievable target for a newspaper press. Be safe: use a preferred prepress vendor who's worked with newspapers before.

You want your 4-color newspaper ad to look great? Take the time up front to prepare your materials well.

Chapter 9

FINISHING TOUCHES: AFTER YOU PRINT

In printing, the various finishing processes are like wardrobe accessories—boy, what a difference they can make! Whether you add varnish to that brochure, or decide to foil stamp your book covers, know that you have many options to jazz up every print job. That's where the fun begins. Here's a look at some of those accessories.

Varnish: More Than a Garnish

Like an exotic spice that transforms a Sunday night stew into a memorable "piece de resistance," varnish can dramatically enhance the beauty of a printed piece.

Many people think varnish is only used to protect printed materials from scuffing or fingerprinting. But it's also used to create a dramatic dimension in a graphic design.

There are three main types of varnish: gloss, dull, and satin. Each imparts a different feel, or effect, to the finished piece. Different types can even be used in tandem or in juxtaposition.

Varnish is basically a type of ink. It's usually clear, although it can be tinted. You can apply overall or "flood" varnish on the entire sheet, or you can apply varnish on specific areas. This is called "spot" varnish. Printers apply varnish both in-line (on press) and off-line (after the printing process), on sheet-fed offset and web offset presses. It works best on coated stock.

Varnish can make images pop out from the page or retreat subtly into the background. It can intensify contrasts, highlight color, and create illusion.

You can't suddenly decide to varnish your job at the last minute. Work closely with your designer and consult your printer before specifying it, as varnish is influenced by paper as well as inks.

And get that tired old image of a glossy, varnished annual report cover out of your mind. Varnish is much more than a protector. Let it add some glamour to your next piece. It's subtle and powerful at the same time.

The Common Fold

There's an art to folding paper. Let's talk about origami.

Just kidding. We're going to look at common folds for print jobs. Even the simplest folds require planning. If you read no further than this paragraph, remember two things: 1) Common folds have standard names, so use these names when talking with printers and designers; and 2) provide your printer with a folding dummy. Printers aren't mind readers.

It all boils down to two basic ways to fold paper: parallel folds and right-angle folds. A parallel fold runs parallel to a previous fold. Think of folding a letter—two parallel folds are used to fit a letter into a #10 envelope. Right-angle folds (there have to be at least two) are at 90 degrees to each other.

Common types of folders are the 4-page, the 8-page, the

12-page, and yes, the 16-page folder. These are created by various combinations of parallel and right-angle folds. Have you heard of the French fold?

Picture a formal wedding invitation; it's an 8-pager created by two folds at right angles to one another. Fold a sheet of paper in half, then fold it in half again, and voila, le French fold.

Letterfolds are probably the most common. Remember to tell your printer if you want your letter flat or folded. If the address on the letter will show through a window envelope, ask to have the piece folded "copy out" and provide a dummy, too.

Then you have your accordian folds, or fan folds. They're simple zig-zag folds that go in opposite directions. Picture the letter "Z." That's the accordian fold.

Gate folds are also known as window folds, but I don't believe Bill Gates and his Windows OS have ANYTHING to do with them. Gate folds are formed by folding left and right edges of the paper inward with two parallel folds. The folds meet smack in the middle, no overlapping. Picture a window with shutters. Now close those shutters over the window, and you have a gate fold. Think of it as the "peek-a-boo fold."

Remember that folding isn't totally precise. So if your design's impact is hinging on a perfect fold (can't resist a bad pun), talk with your printer about what's realistic to expect.

Speaking of design, be sure your designer takes folds into consideration when he or she is doing your layout. Final folds impact copy and image position.

Are we done yet? Nope. Consider your paper, too, for a couple of reasons. First, paper grain affects foldability. Paper always folds more easily in the direction of the grain. Going with the grain (as opposed to against it) also helps to avoid paper cracking.

Paper weight also matters. If paper's too heavy, letterpress scoring may be required before folding, to ensure an even fold. Paper that's too light may be difficult to feed into the printer's folding machine.

Finally, think about mailing issues, particularly with fancy folds. Some mail processing machines may not be able to handle certain pieces. Don't wait till your piece is printed, folded and ready to be mailed. Think about mailing issues in the planning stages. And if your piece gets inserted into envelopes, determine early on if you can use standard envelopes vs. custom made ones.

There's a lot to consider, isn't there? These are just a few of the common folds and related issues to keep in mind. I take back my opening sentence. Folding for printing isn't so much an art as it is a science.

Talk it up. Do it right.

Two Lasting Impressions: Stamping & Embossing

Finishing processes can add texture and depth to your printed piece. Foil stamping and embossing are two popular finishing techniques. If you've admired annual report covers, pocket folders, stationery products, announcement cards, calendars, or book jackets, chances are that your eye is being drawn to the finishing touch, which just might be foil stamped or embossed.

I interviewed Jay Smith, President of Matheson Higgins/ Congress Press (www.mhcp.com.) in Woburn, Massachusetts, on these two subjects.

Finishers are not commercial printers; they're specialists who take printed materials from a printer and perform one of several processes to give them additional impact and beauty. As Jay's Web site states, "We gild the lily!" If your print job requires foil stamping or embossing, your printer will work with a graphic arts finisher.

Sometimes called leaf stamping, foil stamping is the process of applying a pigment to paper or plastic. Most of the pigments (and there are hundreds of colors and patterns) are of bright metallic finish, but there are also satin metallics, gloss and matte pigments, pastels, pearls—and even holographic foils. The foils, or pigments, come on rolls and are applied to the sheet of paper on a stamping press using a die, heat, and pressure.

Are there many kinds of foil stamping, I wondered? A flat foil stamp is the most common, Jay noted, although there's also a combination stamp, in which the foiled image is raised.

The embossing process, on the other hand, always adds dimension, since it raises an image on the surface of the paper. (You can do the opposite, too, called a deboss.)

Both stamping and embossing are done on presses that are very different from offset printing presses. The image to be stamped or embossed is transferred to a die, which is heated on press at temperatures ranging from 200 to 350 degrees. The die comes in contact with the paper at 10 to 150 tons of pressure.

Because so much pressure is being applied to a thin material, it's critical that the die hits the paper evenly in all areas. Too much pressure in stamping will leave an impression on the back of the sheet. And too much pressure in embossing will cause the paper to crack. It's the job of the press person to apply just the right amount of pressure.

Blind embossing creates a raised impression on a blank sheet of paper, not over a printed image, although you can emboss over a printed image, which is harder to do and more expensive.

Many people think that there's only one level of embossing, when in fact, there are many, as Jay explained. A single-level emboss, the most common, raises the whole image to the same level. A multi-level emboss raises the image to distinct multiple levels (usually two to four). Picture your corporate logo inside a box, where the logo's embossed at one level and the box at another. That's a multi-level emboss.

There's more! Sculptured or modeled embossing has many, many levels, but none are distinct. One example would be embossing a person's face or a building. These embossing dies are modeled by hand and are the most expensive.

You should also know that there are many different types of edges for an embossing die, including rounded, beveled, or straight.

Is paper choice critical? Absolutely! Certain stocks limit the depth of the emboss or the sharpness of the foil edges. Jay says that coated stocks are great for stamping. Their smooth surface allows the foil to break cleanly, producing crisp edges and the ability to hold fine detail. But the results are different when embossing. Because of a coated stock's hard clay surface, the embossing may crack or break through.

The stock's thickness also plays a key role. The thicker the stock, the deeper the emboss, and the less likely the image will show through the back of the sheet when stamping as well. Jay's firm likes to have a sheet of the actual stock when making a die for a customer.

You can see a proof of stamping or embossing before your job is actually run. It can get expensive if the finisher sets up your die on the press to make a proof. An alternative is to stamp or emboss the actual paper with a die that's currently on a press (someone else's die), so that you can see how the foil or embossing will look. Usually there's no charge to do this. Then when your job runs, you should do a press check.

Are there common mistakes that consumers make when specifying stamping or embossing? "I think the biggest problem is when we are unable to meet their expectations," Jay noted. "When we're involved early in the process, it really helps eliminate any problems."

Be sure to show your finisher your layout and design when requesting estimates. Finishing processes, like everything else in printing, are not "push-button simple." They're complex specialties. You're creating something beautiful, so meet the builder early!

Thermography: Raised Printing

Many people are misinformed about thermography.

Commonly called raised printing, thermography is also referred to as "poor man's engraving" because it costs significantly less than engraving and looks and feels very similar to the untrained eye.

Thermography is a multi-step process. After sheets of paper are printed on an offset press, they pass under the thermography unit, which sprays colorless powder on them. The powder sticks to the wet ink, and a vacuum sucks up the excess powder. The remaining powder is melted into the ink by a heating unit, causing it to swell up as it melts, and rise above the paper. Voila! Raised printing!

The thermography takes on the color of the ink below it—so you can get thermography in any color.

Some but not all printers charge more for thermography, so shop around for price as well as quality. Also, make sure you're promised laser-safe products. Your letterhead and envelopes need to work with your laser printer.

P.S. Thermography can also be effective for book covers, announcements, and invitations.

Engraving: Printing of the Finest Kind

When I think of engraving, the phrase "the finest printing" comes to mind for two reasons. First, I picture elegant letterhead, formal certificates, dazzling annual report covers, and U.S. currency, among other engraved samples. Second, it's a fact that engraving produces the finest or sharpest image of any printing process.

How do they do it, and what do you need to know about engraving if you're a consumer?

Being the Nancy Drew of printing, I like to follow experts around as I uncover the facts for a topic. For this Tip, I visited MassEnvelopePlus in Somerville, Massachusetts (www.massenvplus.com). Contrary to their name, they don't just do envelopes. My host was Mike Dubie, Vice President of Sales & Service Manager.

What makes engraving so fine? Simply this: images are chemically etched or hand carved into the surface of metal plates, called dies. Once positioned on the press, the engraved die is totally covered with ink and then wiped smooth. This leaves the ink inside the engraved image. Extreme pressure is applied to the paper (approximately 2000 lbs. per square inch) as it's fed into the press and over the die, forcing the ink into the etched image.

The finished result is sharp, crisp, and distinctive: you can always feel an engraved image. Just run your fingers over it from either side. The engraved image is three-dimensional. And you'll feel the indentations of the image from the other side of the sheet.

So its beauty is tactile as well as visual.

There are different kinds of engraving dies: steel dies and copper dies. If you were engraving 100,000 pieces of letterhead, as opposed to 1000, your engraver would use either a steel die or a copper die that's chromed (adding chrome makes the die last longer).

Although many dies are still cut by hand (known as hand-tooled), many images can be photo-etched from negatives created from digital files.

There are different styles of engraving dies, depending on the specific image you want. You can have different depths to your image and different kinds of edges beveled into it.

Engraving is done on special engraving presses, some of which date back to the early 1900s. (The process itself dates back to the 1400s.) The old handfed presses (a beauty to behold, if you love printing equipment as I do) can engrave 500 to 1000 pieces an hour. Newer engraving presses are automated and can produce more than 4000 pieces an hour.

Engraving inks are special, too. They're water based, not oil-based or soy-based as are other inks. They're also totally opaque, not transparent. This gives customers added flexibility, because you can engrave light colors on dark backgrounds.

Another benefit to engraving is that there is zero variation in color throughout the run. "The color's dead-on every time," said Dubie.

Metallic engraving inks have actual flakes of metal (like silver or copper) mixed into them. For an extra special metallic effect, the engraved image can be burnished, giving it a smooth, buffed look.

There are some size limitations with engraving. The largest image area that can be engraved on one die is 5" x 9." If your job exceeds this area, you'll have two dies made. For example, if images at both the top and bottom of your letterhead need to be engraved, two dies will be made to accommodate them.

Engraving is laser compatible, because the ink doesn't smear or "break down" when subjected to the heat of desktop printers. (Be safe and double check this with your printer.)

Need designers prepare their files any differently for engraving? Not at all. Dubie recommends, however, that customers contact a professional engraver early on. Engraving is often combined with other processes, including offset printing, embossing, foil stamping and die cutting. It's usually the final step in the production process, making it critical for customers to plan the job with the engraver in mind. For one thing, offset inks need to be wax-free if you're going to having a printed piece engraved.

You might think that engraving is too costly, but Dubie

notes that it's not as expensive as some people think. Another fallacy is that engraving takes forever to do: he can get a die created in two days and turn a job around in five to seven days.

Engraving is a fine old art, not a lost art, and a specialty not done by most printers. There are many applications for "the finest printing," so when you need to convey an image of success and high-quality, consider engraving.

If you have questions, you can contact Mike Dubie of MassEnvelopePlus at 800-368-1368 or send him an email at mdubie@massenvplus.com.

The Ties That Bind

Printers often do more than printing—they do binding, too. After all, many printed materials need to be bound. And many need to be mailed.

There are lots of different ways to bind your printed pages together. Here are the most widely used ones:

Saddle-stitch (or saddle-wire) binding is the most common method for binding booklets, calendars, and many magazines, such as Time. You've seen it a million times. Two wire staples are machine-inserted through the spine (or back) of the booklet. Your booklet can lie flat. You probably don't even notice the staples.

Paste binding is an alternative to the above. There's no metal involved, only glue, which is applied along the fold lines of the paper. It's cheaper than saddle stitching, but can only be done on lightweight paper. (When I bought financial printing, I used it for my firm's web-printed prospectuses, saving tons of money.)

Side stitching means the two staples are put through the entire stack of printed pages from front to back. Picture two staples along the left side of your booklet. It's not elegant, and your pages can't lie flat. Some copier machines can do side stitching.

Two popular mechanical binding methods are **Wire-O** and **GBC**, both known by the manufacturer's names. Holes are punched in the paper, then either spirals or combs made of plastic or wire are inserted. It's common with software manuals and many cookbooks. Unit costs are high.

Paperback books and phone books are **perfect bound.** Gathered sections of a book, called signatures, are aligned at the spine, trimmed at the edges, and then glued. The cover is then pressed against the spine, and the book is trimmed again on the other 3 sides. Perfect binding costs more than saddle stitching.

Hardbound books are **case bound** (also known as edition binding). The sections of pages (signatures) are sewn together along the spine, gathered, and trimmed. The covers, or cases, are pieces of cardboard covered in paper, cloth, or other materials. They're imprinted and then pasted onto the insides. Case binding costs the most but lasts the longest. Many steps are involved, and only certain binderies can do it.

These are only the most common binding methods. Most commercial printers offer saddle stitching and will outsource methods like perfect binding. Talk with your printer about binding options as you plan your job.

"... the Envelope, Please!"

There's more to envelopes than meets the eye. To give consumers a broad overview of envelopes, I spoke with Linc Spaulding, VP of Sales for Sheppard Envelope in Worcester, Massachusetts, a specialty envelope converter (www.sheppardenvelope.com).

Here's what Linc had to say about envelopes:

When buying envelopes, you get what you pay for—and prices are all over the map. It pays to shop, and knowing where to shop can be critical.

By far the least expensive envelope for its size is a #10 (4 1/8" x 9 1/2"). It's the perfect size for mailing business correspondence, because your standard 8 1/2" x 11" letter fits easily inside when folded twice. All envelope-inserting machines can handle #10s, and they easily hold a business reply envelope, also known as a #9. Of the billions of envelopes manufactured every year in the U.S., more than 35% are #10s. They're the ultimate envelope commodity and are priced accordingly. Office supply stores are great places to buy boxes of 500 or cartons of 2,500.

If you're doing direct mail and you don't want your mailing to end up in the trash, try to make the envelope distinctive, either with a design or with a specialty paper. It's key to getting a good response rate. (Consider including a reply envelope as well, since the increased cost on the total mailing is negligible.)

Over the years several hybrid envelope/form combinations have been developed that essentially make a complete direct mailing out of a single sheet of paper. Not only are these formats environmentally friendly, but they can also cut costs substantially. They're certainly worthwhile if you are interested in minimizing the costs.

There are fewer than 80 envelope plant owners in the U.S., many of whom own more than one plant. Due to a trend toward consolidation, envelopes have become even more of a commodity. Finding a plant to manufacture a custom envelope in smaller quantities (up to 25,000) is increasingly difficult.

Look up "envelopes" in the Yellow Pages and you'll find countless listings. It's impossible to tell which firms are manufacturers and which are simply brokers (or jobbers) with minimal printing capacity or jobbers with pretty good printing capability but no folding capability.

Keep in mind when designing your envelope that it's important to find the right source. The more sophisticated the printing, the more exotic the stock, the more difficult, time-consuming and expensive it will be to put the package together. Find a manufacturer who likes to work with commercial printers, or a commercial printer who's comfortable working with envelope conversions.

On the other hand, if all you want is plain vanilla #10s or 9 x 12" booklet envelopes printed in one or two colors, make your decision based on price and delivery. There are tons of "envelope printers" (distinct from commercial printers) who can accommodate you.

P.S. Don't overlook the many detailed postal regulations about designing/printing envelopes. A good resource is *http://pe.usps.gov*.

INDEX